The Urban Book Series

Aims and Scope

The Urban Book Series is a resource for urban studies and geography research worldwide. It provides a unique and innovative resource for the latest developments in the field, nurturing a comprehensive and encompassing publication venue for urban studies, urban geography, planning and regional development.

The series publishes peer-reviewed volumes related to urbanization, sustainability, urban environments, sustainable urbanism, governance, globalization, urban and sustainable development, spatial and area studies, urban management, urban infrastructure, urban dynamics, green cities and urban landscapes. It also invites research which documents urbanization processes and urban dynamics on a national, regional and local level, welcoming case studies, as well as comparative and applied research.

The series will appeal to urbanists, geographers, planners, engineers, architects, policy makers, and to all of those interested in a wide-ranging overview of contemporary urban studies and innovations in the field. It accepts monographs, edited volumes and textbooks.

More information about this series at http://www.springer.com/series/14773

Donald Okeke

Integrated Productivity in Urban Africa

Introducing the Neo-Mercantile Planning Theory

 Springer

Donald Okeke
African Settlements Research Group,
   Department of Urban and Regional
   Planning
University of Nigeria
Enugu, Enugu State
Nigeria

ISSN 2365-757X   ISSN 2365-7588 (electronic)
The Urban Book Series
ISBN 978-3-319-41829-2   ISBN 978-3-319-41830-8 (eBook)
DOI 10.1007/978-3-319-41830-8

Library of Congress Control Number: 2016945967

© Springer International Publishing Switzerland 2016

This work is subject to copyright. All rights are reserved by the Publisher, whether the whole or part of the material is concerned, specifically the rights of translation, reprinting, reuse of illustrations, recitation, broadcasting, reproduction on microfilms or in any other physical way, and transmission or information storage and retrieval, electronic adaptation, computer software, or by similar or dissimilar methodology now known or hereafter developed.

The use of general descriptive names, registered names, trademarks, service marks, etc. in this publication does not imply, even in the absence of a specific statement, that such names are exempt from the relevant protective laws and regulations and therefore free for general use.

The publisher, the authors and the editors are safe to assume that the advice and information in this book are believed to be true and accurate at the date of publication. Neither the publisher nor the authors or the editors give a warranty, express or implied, with respect to the material contained herein or for any errors or omissions that may have been made.

Printed on acid-free paper

This Springer imprint is published by Springer Nature
The registered company is Springer International Publishing AG Switzerland

Foreword

In this book, *Integrated Productivity in Urban Africa: Introducing the Neo-Mercantile Planning Theory*, the author endeavours to and successfully makes a valuable contribution to the theme by introducing neo-mercantile planning theory to the development of spatial systems. Consequently, this unique treatise focuses on those components that direct integrated productivity within the urban components of Africa spatial systems. The book addresses the resilient gap in existing knowledge and related approaches through application of planning instruments as a necessary measure in redressing the urban productivity question typical of the African economic systems.

The contents of the book is well structured in terms of the reigning perceptions of urban productivity in Africa and the building blocks consisting of definitions, terminology, methodology, analysis and conclusions. The book includes a critical assessment of the neoliberal philosophy and its interface with related theories from both urban and regional perspectives.

Both academics and practitioners share the need to formally address the challenge of integrated productivity in the African spatial context through the adoption of applicable planning theory based on neoliberal influences and developments. Dr. Okeke points out that the neoliberal planning serves as the delivery instrument within neoliberal economies. In focusing on the failure of neo-liberal planning theory and practice deriving mainly from inherent vitiation of green growth and spatial justice in Africa, the critical thinking and analytical abilities of the writer are brought to bear.

In particular, the writer addresses compliances to neoliberal participatory planning within certain national spatial systems of South Africa, Egypt, Ethiopia and Tanzania and its implications for the transformation in the education systems. It is essential to rethink the role of current education, training and practice, if sustainable development in both economic space and spatial development is the focus.

The book proposed a new direction in the application of neo-mercantilism, and the way its related theory and planning instruments is developed. It rethinks the traditional planning and development orthodoxies and approaches in terms of an alternative theory with supporting territorial planning framework instruments to promote spatial integration. It establishes the development of alternative spatial planning theory within Africa, a continent that is historically and otherwise being dominated by the influences and impacts of colonial legacies as far as spatial planning and development are concerned. This book contributes to the development of international theory in line with existing spatial, economic and political realities on the African Continent. I believe the work is a ground-breaking publication as it concerns the understanding of urban spatial systems and planning in Africa.

Prof. Carel B. Schoeman
Unit for Environmental Sciences and Management
NWU (Potchefstroom Campus), South Africa

Preface

The relevance of this book draws from the engagement with sourcing new Southern perspectives on urban and regional planning. For a long time and until now, there exists a resilient gap in knowledge regarding the determination of appropriate planning instruments for redressing urban productivity decline which has plagued African economy. The intellectual community has contended with this challenge in the context of global hegemonic influences on planning theory. The dynamics in planning theory exacerbated with the ascendance of neoliberalism as global economic orthodoxy since 1980s. This reality unleashed changes epitomized in neoliberal planning theory. Neoliberal planning serves as delivery instrument in neoliberal political economy. After more than three decades of experimentation, neoliberal planning theory is called to serious question due to indications of its poor performance in the delivery of green growth and spatial justice in Africa.

Regardless the generally appreciable level of compliance to neoliberal participatory planning and the above average compliance identified for some African countries specifically South Africa, Egypt, Ethiopia and Tanzania, the move for reforms in the curriculum of planning education in favour of neoliberal planning model is literally put on hold following recent recourse to rethink neoliberal participatory planning paradigm. The rethink, which is intended to reinforce the theoretical arsenal of planning, informed the sub-theme of South-South engagements: new Southern perspectives on urban planning in the 2014 AAPS conference. The forthcoming 2016 World Planning Schools Congress: Global Crisis, Planning & Challenges to Spatial Justice in the North and the South, organized by the Global Planning Education Association Network (GPEAN), is also concerned with exploring Southern perspectives for planning.

The content of the proposed book is a timely response that postulates a theoretically compelling Southern perspective of planning in the African context. Given the precarious disposition of African political economy and planning experience, the book proposes a new direction, which rethinks neoliberalism as meta-theory of planning in Africa. It pioneers an original school of thought that presents a general theory of planning for Africa in the twenty-first century.

The book analyzes the development of planning paradigm in Africa since the turn of the twenty-first century. The primary purpose is to provide an insightful rethink of trends in the development of existing planning orthodoxies. These planning orthodoxies, which were conceived in the 1980s and thereafter with the mindset that upholds neoliberal values, are diagnosed to have difficulties with delivering spatial integration in Africa. Thus the book is particularly curious about the vogue of informality in planning and argues that it derives from the meta-theory that subjects Africa to dependent capitalism. The book contributes an alternative planning theory along with territorial planning framework instruments that are sensitive to the delivery of spatial regional integration for African renaissance.

The wide range of end users anticipated, which embrace the political class, the intellectual elite, and the academia who are concerned with planning, will find the book useful at the meta-theoretical, theoretical and practical levels. At the meta-theoretical level, neoliberalism as development ideology for Africa is called to question. This is done against the backdrop of concerns for epistemological foundations of external domination and ideologies in African political economy. The re-engineering of imperial trends is topical for new regionalism which currently engages the attention of the international community—the political class and the intellectual elite especially political scientists and role players at the African Union (AU). The theoretical level discusses the dynamics in the science of planning, its links and relationships with meta-theoretical issues, and the contributions of the academia and research institutes as primary role players in this arena. Thus the book contributes planning theory reforms as an entry point to impact the pending review of academic curriculum for planning education in Africa. Then at the practical level the book serves as food for thought for the intellectual community and as point of reference for planning consultants and development agencies.

Enugu, Nigeria   Donald Okeke

Acknowledgments

May I acknowledge the divine providence which propelled the delivery of this work and the team spirit of my wife, Agatha and children, Paschal, Anita-Christie, Audry-Rose, and JohnPaul; also my elder brother Eric for his inspirational talks. I am grateful to Prof. C.B. Schoeman, my study leader at North West University (NWU), who inspired my disposition as African region scholar in planning studies. His intellectual guidance and that of my co-study leader Dr. E.J. Cilliers in the delivery of my research at NWU, which informs the major contribution of this book, is herewith acknowledged. In the same vein, I acknowledge Prof. H.S. Steyn of the Statistical Consultation Unit, NWU for his practical involvement in my research work. I accept responsibility for any omissions and errors.

Contents

1	**Introduction**		1
	1.1 Perceptions of Urban Productivity in Africa		1
	1.2 Extended Definition of Terms		6
		1.2.1 Urban Environment	6
		1.2.2 Spatial Planning	8
		1.2.3 Space Economy	10
		1.2.4 Imperialism	12
		1.2.5 Neoliberalism	15
	1.3 Methodology		16
	1.4 Analytical Approach		17
	1.5 Synthesis		18
	1.6 Structure of Book Report		19
	1.7 Conclusion		19
	References		20

Part I Theoretical Founding

2	**Points of Departure and Major Arguments**		25
	2.1 Points of Departure		25
	2.2 Assumptions		27
	2.3 Major Arguments		28
		2.3.1 Theoretical Perspective	28
		2.3.2 Analytical Perspective	30
	2.4 Conclusion		31
	References		31

Part II Application of Theory to the Reality Within Africa

3	**Epistemological Ideologies and Planning (in Africa)**		35
	3.1 Imperial Space Economy		35
	3.2 Informality in Planning		38
		3.2.1 Notions of Informality	38

		3.2.2	Notions of Planning.	41
		3.2.3	Informality and Planning: Notions of Relationships.	43
	3.3	Informalization of Cities in Africa		45
		3.3.1	Contextualizing Major Arguments and Presumptions	45
		3.3.2	Diagnosis of Growth Potentials.	47
	3.4	Basic Scenario of Planning in Select African Countries		50
	3.5	Conclusion		77
	References.			77
4	**Meta-Theoretical Frameworks of (Spatial) Planning**			81
	4.1	Traditional Perspective		81
	4.2	International Perspective.		84
	4.3	Neoliberalism as Meta-Theory of Planning.		88
	4.4	Conclusion		91
	References.			92
5	**Theoretical Frameworks**.			95
	5.1	World-Systems Theory.		95
	5.2	Growth of Urban (Structure) Theory.		97
	5.3	Growth of Regional Development Theory		101
		5.3.1	The Concept of the Region and Regional (Economic) Integration	101
		5.3.2	Review of Regional Development Theories (from 1930s)	103
	5.4	Growth of Urban Planning Theory (from 1960s)		110
		5.4.1	Master Planning Paradigm.	112
		5.4.2	The Participatory Planning Paradigm.	113
		5.4.3	Theoretical Framework of Planning.	117
	5.5	Spatial Models for Regional Integration		117
		5.5.1	The Extended Metropolitan Region (EMR) and Growth Triangle (GT) Model	120
		5.5.2	The Poly-centric Model Compared with the EMR Model	121
	5.6	Conclusion		122
	References.			122
6	**Overview of Urban Planning Principles and Practice**.			125
	6.1	Dynamics of Paradigm Shift in Planning Perspective		125
	6.2	Trends in Statutory Planning Perspective		126
	6.3	Trends in Spatial Planning Perspective.		127
	6.4	Trends in Spatial (Urban) Planning Initiatives		130
		6.4.1	Review of Sustainable Cities Programme (in Nigeria)	130

	6.5	Trends in Planning Practice: A Case Study of the Resilience of Formal Planning Practice	136
		6.5.1 Methodology	137
		6.5.2 Preliminary Literature (Theoretical and Analytical Frameworks)	138
	6.6	Summary of Basic Scenario of Planning Practice	142
	6.7	Conclusion	145
	References	146
7	**Analysis of Current Reality (Principles and Practice)**	149
	7.1	Functional Flow of Analyses	149
	7.2	Deductions of Analysis	151
		7.2.1 Timeframe Analysis	151
		7.2.2 Regional Analysis	152
		7.2.3 MCA Analyses	152
		7.2.4 Perception Analyses of Desktop Case Studies	156
		7.2.5 SWOT Analyses/Own Assessment	158
	7.3	Test of Hypothesis	160
	7.4	Scenario Analysis	162
	7.5	Conclusion	164
	References	165

Part III Synthesis and Statement of New Theory

8	**African Renaissance**	169
	8.1	African Renaissance—a Theory of Development	169
	8.2	Conclusion	177
	References	178
9	**Neo-Mercantilism as Development Ideology (in Africa)**	181
	9.1	Neo-Mercantilism: A Rationale for Change	181
	9.2	Neo-Mercantilism as Development Ideology (for Africa)	184
	9.3	Neo-Mercantile Versus Neoliberal Development Ideologies	188
	9.4	Conclusion	190
	References	191
10	**Introducing Neo-mercantile Planning Theory**	193
	10.1	Neo-African Spatial Development Theory	193
		10.1.1 Major Argument of Neo-African Spatial Development Theory	196
		10.1.2 Neo-mercantile Planning Concept	198
		10.1.3 Neo-mercantile Planning Paradigm	200
		10.1.4 Neo-mercantile Planning Theoretical Framework	202
		10.1.5 Elements of Neo-mercantile Planning Theory	203

10.2	Spatial Integration Network	214
10.3	Urban Spatial Model	219
10.4	Measure of Time-Efficient Coefficient for the Classification of Settlements	221
10.5	Conclusions	222
References		223

Part IV Contribution to New Knowledge, Application and Approaches

11 New Knowledge in Planning (in Africa) ... 227

11.1	Visioning Process for Spatial Integration Network in Africa		227
	11.1.1	Priority Problems	228
	11.1.2	Vision Exposition	228
	11.1.3	Vision Objectives	229
11.2	Priority Actions		231
	11.2.1	Assessment of Current Actions	231
	11.2.2	Priority for Action	235
	11.2.3	Proposed Action Cards	235
	11.2.4	Typology of Action	235
11.3	Implementation Strategies		238
	11.3.1	Institutional Requirements and Implementation Processes	239
	11.3.2	Manpower Requirements	242
	11.3.3	Financial Mechanisms	242
	11.3.4	Legal Reforms	242
	11.3.5	Monitoring Measures	243
11.4	Calendar of the Action Plans for Spatial Regional Integration in Africa		243
11.5	Conclusion		245
References			246

Appendix A ... 247

Glossary ... 257

Index ... 261

Abbreviations

AAP	African Association of Planners
AAPS	African Association of Planning Schools
ADB/ADF	African Development Bank
AGOA	Africa Growth and Opportunity Act
AMCHUD	African ministerial conference on housing and urban development
APRM	African peer-review mechanism
ASDP	African spatial development perspective
ASEAN	Association of South East Asian Nations
AU	African Union
CAADP	Comprehensive Africa Agricultural Development Programme
CAP	Commonwealth Association of Planners
CBA	Community-based approach
CBD	Central Business District
CCD	Convention to Combat Desertification
CDS	City development strategy
CEMAT	European Conference of Ministers Responsible for Regional Planning
CEN-SAD	Community of Sahel-Saharan States
CODESRIA	Council for the Development of Social Science Research in Africa
COMESA	Common Market for Eastern and Southern Africa
DCs	Development corridors
DRC	Democratic Republic of Congo
EAP	Environmental Action Plan
ECA	Economic Commission for Africa
ECCAS	Economic Community of Central Africa States
ECOWAS	Economic Community of West African States
EMR	Extended Metropolitan Region
EPM	Environmental Planning and Management
ESDP	European spatial development perspective
EU	European Union

FDI	Foreign direct investment
GDP	Gross domestic product
GDS	Growth and Development Strategy
GIS	Geographic information system
GT	Growth Triangle
HOS	Heckscher–Ohlin–Samuelson
ICT	Information and communication technology
IDP	Integrated development planning
ILO	International Labour Organization
IMF	International Monetary Fund
IPPP	Integrated physical planning approach
IRDA	Integrated Regional Development Act
LEED	Local Economic Empowerment and Development
MCA	Multi-criteria analysis
MDG	Millennium development goal
MPA	Master planning approach
NAIRU	Non-accelerating inflation rate of unemployment
NBG	NEPAD Business Group
NEPAD	New perspectives on Africa development
NIRDA	National Integrated Regional Development Act
NSDP	National spatial development perspective
NWU	North West University
O&OD	Opportunities and obstacles to development
OAU	Organization of African Unity
PSDS	Spatial development strategy/framework
PUDs	Planned unit developments
R&D	Research and development
RAIDS	Resource-based African industrial and development strategy
RIFF	Regional Integration Facilitation Forum
SADC	Southern African Development Community
SAP	Structural Adjustment Programme
SCP	Sustainable Cities Programme
SDF	Spatial development frameworks
SDIs	Spatial development initiatives
SEED	State Economic Empowerment and Development
SEP	Sustainable Enugu Project
SEPM	Strategic environmental planning and management
SIP	Sustainable Ibadan Project
SIPA	Spatial integration planning approach
SIPs	Spatial integration planning
SKP	Sustainable Kano Project
SMEs	Small and medium enterprises
SUDP	Strategic Urban Development Plan
SWOT	Strength, weakness, opportunity and threat
TIPs	Thematic Integration Planning

UDF	Urban development frameworks
UGBs	Urban growth boundaries
UK	United Kingdom
UMP	Urban Management Programme
UN	United Nations
UNCHS	United Nations Center for Human Settlement
UNDAF	United Nation's Development Assistance Framework
UNDP	United Nation's Development Programme
UNEP	United Nation's Environmental Programme
USA	United State of America
USAID	United States Agency for International Development
WAEMU	West African Economic and Monetary Union
WHO	World Health Organization

Chapter 1
Introduction

> *Discovery consists of seeing what everybody has seen and thinking what nobody has thought ......SIFE.*

Abstract The resilience of the epistemological foundations of imperialism is diagnosed to be the problem with urban productivity in Africa. Planning responses fail to address this malaise because planning scholarship barely extends beyond the consideration of the science of planning. The meta-theory of planning, which engages the influence of development ideologies on planning, is seldom considered. This omission disenfranchises planning in the chessboard of political economy. At the moment, planning in neoliberal dispensation is delivering dependency in which spatial systems in Africa function as imperial instruments. An extended definition of terms is given in anticipation of the launch of the new planning theory.

Keywords Imperialism · Epistemology · Resilience · Formal planning · Neoliberal planning · Productivity · Informalization

1.1 Perceptions of Urban Productivity in Africa

Urban productivity decline is a common feature associated with African economy. It has been severally related to urbanization without growth. James Hicks of the World Bank gave a graphic representation of this reality in 1998 and argued that over the past 35 years the root cause of this phenomenon is not known (Hicks 1998). In establishing his position, he indicated that over the 1970–1995 period, the average African country's urban population grew by 5.2 % per annum while its' GDP declined by 0.66 % per year. This scenario has not significantly changed. What obtains in recent times is 'growth without development' still underlain with declining productivity. That is to say, growth expressed in positive GDP diagnosed in the context of increasing poverty and unemployment. Increase in investment, especially through FDI, seem not to generate transformative growth of social systems in Africa. The perplexing growth which occurs has engaged the attention of the international community over the years. Although since the turn of the twentieth century, the concern of the international community is more out of fear that the declining African economy may adversely impact the global economy and not sympathy for Africa.

There are several perceptions of the performance of African economy. Perception approaches varies and with reference to the dominant informal economy in Africa Owusu (2007) diagnosed three positions: reformist, institutionalist and neo-Marxist. At the moment institutionalist perspectives are very influential in policy circles and have been incorporated into the work of neoliberal economists, policy advisors and non-governmental organizations, partly because it conforms to the global push for neoliberal and supply-side economics (World Bank 1989: 10). The reformist perspective shares most of the indifference of the institutionalist position towards exploitative capitalist economy in which Africa is at the down side. On the contrary, the neo-Marxist position is curious about the exploitative relationships in capitalist production and distribution. The world system driven by globalization subjugates this position in favour of institutionalist perceptions which is receptive to the epistemologies that retain Africa in dependency.

Given the dominance of institutionalist perception, urban predicaments have increasingly ceased to be considered in the context of the epistemologies of African civilization. Although in the psyche of imperialists who endorse external domination, African civilization is a nullity. This impression, which is expressed implicitly in literature, has the tendency of signalling the dearth of an annals approach in diagnostic research for African development. What often obtains is spot evaluations fraught with banal diagnoses of causalities in African urban dynamics. It is not uncommon to notice symptoms of this lapse tucked in here and there in accounts of urban planning in Africa as presented by some African scholars, particularly those commissioned by external assistance agencies. Most of these studies are related to UN research efforts to transcribe urban planning in Africa. To this end attention is drawn to three topical studies, namely, reassessment of urban planning in African cities in 1999; a regional overview of the status of urban planning and planning practice in Anglophone (sub-Saharan) African countries in 2009; and revisiting urban planning in sub-Saharan Francophone Africa in 2009.

This contribution shares the neo-Marxist concerns that appeal to redressing the submersion of Africa in dependent capitalism. Therefore, it recalls that the African economy is built on epistemological foundations of external domination which reflect in environmental and labour capital exploitation (Nabudere 2003). This is responsible for polarized pattern of development otherwise economic dualism is sometimes equated with inequality. The scenario is known to fundamentally influence the nature of urbanization in Africa and by implication economic growth and productivity. Hicks (1998) speculates that there could be some remaining distortions in Africa's urban economies and urbanization process that need to be identified and rectified. His speculation confirms a possible disconnect between urbanization and economic growth in African countries. Although there is little indication of an underlying theory which ties urbanization and economic growth together. Information available on their relationship such as the economy of scale argument is often verbal and/or analytically speculative, dealing with correlations rather than theoretically compelling causations (Graves and Sexton 1979).

1.1 Perceptions of Urban Productivity in Africa

The African economy is the outcome of major political and socio-economic transformations in the form of colonization and the installation and management of capitalism, which defined historic systems in African civilization since mid-fifteenth century. As Comfort Chukuezi (2010) stated 'the process of capitalist penetration of World economy and subordination of the Third World to the capitalist system is effectively described in the works of authors such as Frank (1971) Bairoch (1975) and Wallerstein (1974)'. Deriving from these works Chukuezi aptly deduced that the African economy, as it is the case with those of other developing countries, is located at the periphery of the international capitalism. These economies are structured to meet the needs of the capitalist system (Roberts 1978; Santos 1979; Lowder 1986). They are made to assume a dependency status through various mechanisms such as the control of commodity markets, the activities of the multinational companies and their subsidiaries, the control of the source of credit and interest rates and siphoning of resources from the indigenous society, etc. (Chukuezi 2010: 131).

This had a profound impact on support structures for productivity in African societies. Notable is the change in mindset from transcendental ends in pursuit of nation building to value systems that portray desperate pursuit of material well-being, the worship of private property and the polarization of wealth and income (Dembele 1998). This is followed with structural changes in the spatial systems particularly the space economy commonly tagged economic geography and the city. The structure of spatial systems and their cities was primarily a result of the manner in which they had been integrated into an international capitalist system from the sixteenth century, and the continuing pattern of integration under peripheral capitalism which is very different from the capitalist mode of production in the developed capitalist countries (Chukuezi 2010: 131). The spatial systems became imperial instruments for the delivery of external interests expressed initially in European mercantilism but now in Euro-American and Chinese mercantilism in the context of prevailing neoliberal global economic orthodoxy.

Therefore, the incidence of imperial space economy in Africa dates back to mid-fifteenth century at the wake of European mercantilism. The colonial experience exacerbated the situation leading to the incidence of imperial colonial towns and the reordering of trade routes. Essentially the economic geography was rewritten to facilitate policy objectives of extraction and exploitative trade relations. Besides the redistribution of trade routes, cities in Africa became hybrids an inevitable product of intervening culture and policy formulation hegemony spurn abroad. African cities no longer derive from indigenous values, attitudes and institutions. They ceased to be responsive to indigenous enterprise and culture. As it were 'African cities' became cities in the Diaspora in their homeland. Indeed indigenous African cities are technically near extinct.

Over the years, the epistemology of external domination remained resilient in the planning and development of the city. Since the turn of the twenty-first century, the informalization of cities is underway courtesy of the receptive attitude towards the informal sector. This scenario is favourably disposed to the persistence of pre-capitalist social formations and indigenous economies that act against the full co-modification of land (in some regions) and labour in the capitalist economy that was introduced (Simone 1998). Already isolated urban hierarchies with limited linkages has developed in the urban region mainly in the form of urban-rural dichotomy and the fragmentation of the private sector with extroverted modern sector sparsely related with the local economy (Hicks 1998). The spatial expression of this trend reveals the incidence of pseudo-urbanization in which tertiary sector rather than productive economic activities stimulate growth.

The informalization process found expression in neoliberal planning tradition which emerged in mid-twentieth century at the hills of severe criticism of classic master planning paradigm. It became the global planning orthodoxy for the delivery of neoliberal global economy in the 1980s. Its basic mechanism implicated paradigm shift in spatial planning perspective from detailed to framework planning. The transition to neoliberal planning theory which suggests the reinvention of planning, received tacit support from UN-Habitat and the works of Watson (2009) which is centred on the doctrine of informality in planning and the works of Roy (Roy 2009) which is centred on the doctrine of informality as epistemology of planning. As a result addressing cross-cutting issues rather than core issues in planning became the vogue. Environmental management and decentralization policies now effectively usurp urban policies. Hence attention drifts from spatio-physical aspects of urban form, expressed in the urbanity of cities, to urban quality issues that dwell on degradation in socio-economic and environmental terms. This explains in part the frail relation between modernist and post-modernist planning.

Consequently, planning practice became increasingly informal and pragmatic and coincidentally in the context of urban productivity decline in Africa. The thematic treatment of planning dominated planning practice thus leading to the emergence of sectoral planning. Sectoral planning caused the subject matter of planning to transit to poverty issues. Thus pro-poor planning received impetus and this heightened the fortunes of informality as a core determinant factor of planning initiatives. The planning initiatives pay lip service to integration because they lack the theoretical bases to address it. The unfolding features underpin the tenets of neoliberal planning theory and independent nations tend to find their own synthesis depending on their local conditions. However, as the tenets of neoliberal planning influence planning initiatives, the resilience of formal planning theory is noticed and remarkably acknowledged by UN-Habitat.

Against this backdrop, current trends in development planning point to the approach of UN-driven new regionalism and the resilient cities concept. This trend is likely to dwarf the attention that sprouted on the concept of spatial justice shortly before the infiltration of the new regionalism and resilient city concept. The tenets of spatial justice do not seem to share the common objective of imperialism in the global logic of restructuring as it is the case with new regionalism and perhaps resilient cities concepts. So far the theoretical base of new regionalism indicates a market-driven concept for managing Euro-American and Chinese mercantilism in neo-imperial terms. This is linked with free-trade relations. Its focus on supporting spatial systems within nation-states links it with the resilience doctrine. This is why the determination of the resilience of cities in Africa calls for conceptual definition otherwise a repeat of the resilience of imperial cities to changes in favour of African renaissance during the independence period in mid-twentieth century is likely to reoccur. So far it is not clear how the resilience doctrine relates with re-engineering the epistemological foundations of imperialism in the development of spatial systems in Africa. In the current syndrome of globalization, resilience to imperialism provides an alternative school of thought for delivering the political economy of African countries. This relates more with the study objectives of spatial justice. Perhaps through this avenue, attempts could be made to rework the epistemologies in spatial systems that retain African economy in dependency.

Unfortunately, the dispositions of neoliberalism and neoliberal planning theory towards informality accede to the disregard of the epistemological foundations of imperialism in the development of spatial systems in Africa. The degeneration of planning into informality consolidates these foundations of imperialism which implicates mindset and development ideology issues, the structure of cities and space economy issues, the orientation of global political economy that schemes to retain Africa in capitalist materialism, the resilience of poverty rooted in dependency theory of the world system, and ultimately exploitative regionalism. These epistemologies, which are by no means exhaustive, are not compatible with enhanced urban productivity in Africa. A change subsumed in renewed planning approach to rework spatial systems is imperative to enhance urban productivity in Africa.

In strategizing for renewed planning approaches, it is imperative that new realities in the outlook of imperialism have to be put into consideration. As such while imperialism is still regarded as the bane of urban productivity in Africa it has since dropped physical coercion in favour of social management to deliver structured coercion of governmentality. Development approaches in contemporary Africa now have to contend with imperialism of administration and imperialism as international administration. The former interprets the system that prevailed in which the emerging ruling class at independence continued acting the script already written by the departing colonizers (Majekodunmi and Adejuwon 2012: 197). This caused economic dependence as epistemology of imperialism to be resilient regardless political independence. The changes in the traditional system of base vis-à-vis culture, value system and worldview remained valid and resilient to reversion. Added to this development are vestiges of imperialism as international administration (Hawksley 2004: 22). This is manifested in the activities of UN-Habitat to aid development in African countries, which tend to turn from support to control (Okeke 2015: 191). This has heightened fears of neo-imperial plot as hidden agenda in hegemonic development ideologies and processes. Nevertheless any renewed effort that will effectively enhance productivity in African countries will have to contend with these realities.

It is conclusive that spatial systems in Africa function as imperial instruments. A change in the functioning of the spatial systems is inevitable and dependent on renewed planning. Some measure of radical approach is required to effect necessary changes that involve direct engagement with demobilizing the new wave of imperialism. To this end this book attempts to introduce a planning theory based on an independent study that analyzed spatial development paradigm for regional integration in Africa. The study conducted in 2010–2015 period in South Africa sets out to review the quiet revolution that is sweeping through planning concepts in the context of new imperialism. To this end, the first wave of growth management in the 1960s and 1970s (Berke 2002: 24) justified the choice of 1960s as a limit to retrospection. The review placed emphasis on African experience to consider the rational for the revolution in planning theory. The study was guided with territorial planning principles along with the mindset that contemplates integration for African renaissance which underpins growth visions for Africa. On account of the study orientation a substantive and not classic planning theory is in contention. The prospective theory portends to derive from convergent hypothesis of endogenous growth as it is established in regional development theories. However, the theory presumes the demobilization of neoliberalism as development ideology in Africa.

1.2 Extended Definition of Terms

The terms isolated for extended definition play systemic role in the diagnosis of planning theory and practice in Africa. This subsection provides a theoretically compelling operative meaning of these terms, in some cases as new constructs of existing knowledge. This is considered necessary to provide a level plying ground and common threshold for subsequent analysis.

1.2.1 Urban Environment

It is common practice to relate urban phenomena to demographic increases and allied demographic indicators such as employment structures. The urban environment is seldom perceived in practice as a dynamic ecosystem that is influenced by natural processes energized with design priorities. This informs the partial understanding of its patterns of growth and change. Subsequently widespread misunderstandings of urban form and the recurrent misdirection of spatial planning are identified. Edwardo (1990: 35) shares this view and Sternberg (cited in Arbury 2005: 59) reinforced it with his argument that urban design lacks a cohesive theoretical foundation. The two approaches of perceiving the urban environment given below are new constructs, which will apply in this discourse.

First, the systems approach illustrated in Fig. 1.1 below upholds the idea that the urban environment is structured in three inter-related layers of natural systems

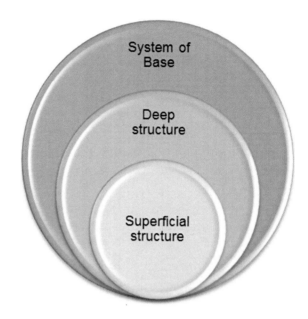

Fig. 1.1 The three layers of urban environment. *Source* Own construction (2011)

comprising the system of base, the deep structure and the superficial structure. The system of base indicates site-morphological factors of culture, technology, environmental factors, value system, world views, cosmology, etc. These factors inform or impact the deep structure which deals with space-activity relationships. The deep structure generates activity systems which are analyzed in different ways to explain land use functions, densities, distributions and growth and change characteristics in the urban environment. The expression of the activity systems in space indicates the superficial structure. This is the spatio-physical form in the visual realm of urban environment that has been widely addressed in theories of urban structure. It represents the expression of the resulting spatial organization of activity systems that characterize the urban environment. A further analysis identifies cultural and spiritual influences at the social-psychological and supernatural realm of the urban environment.

Second, the spatial approach illustrated in Fig. 1.2 below visualizes three spatially segregated sections of the urban environment, namely the urban core area, the inner ring comprising the urban fringe defined as that cultural development that takes place outside the political boundaries of central cities and extends to the areas of predominantly agricultural activity (Arpke 1942), and the outer ring (rural hinterland). The different sections maintain systemic relationships in what constitutes (in spatial terms) the functional region of the urban economy. These sections display functional coherence revealed in the form of the flow of socio-economic activities. It is observed that the spatio-physical distribution of these activities maintain critical relevance in this approach as can be seen in the borderless cities concept of ASEAN countries.

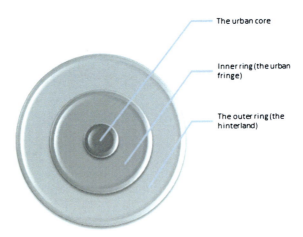

Fig. 1.2 The three segments of the urban environment. *Source* Own construction (2011) derived from literature

1.2.2 Spatial Planning

This aspect of planning concept referred to as 'spatial planning' was apparently taken for granted until the modernist period in planning (i.e. 1970s–1980s) when it drew considerable attention and underwent re-engineering. As the wave of change-gained momentum spatial planning concept started shifting from product-oriented to process-oriented activity, a process that was triggered by heavy criticisms against the master plan instrument. At the outset the master planning critics (Altshuler 1965; Friedmann 1965; Meyerson 1973; Hansen 1968) adopted the technique of diffusing attention from design attributes of spatial planning to socio-economic concerns.

As a theoretical concept, spatial planning diverts attention from the statutory attributes of planning (as represented by statutory planning concept) to the broad spectrum of non-statutory attributes that apply on a wider scale. So in terms of coverage of issues and scale of operation, spatial planning is a more integrated concept. Indeed it is an activity that may take different forms in different contexts, depending on institutional and legal context, or variations in planning cultures and traditions (Healey 1997) Hence spatial planning discourse seldom proffers conceptual definition for spatial planning although Planning Officers Society, 2004 maintain that there are 'many definitions of spatial planning in existence'. There are definitions but often times these definitions are not theoretically compelling. They are either implied or subsumed in assessment statements of the quality of urban development. However, Adams et al. (2006: 45) presented compelling explanations in their analysis of regional development and spatial planning in an enlarged European union. Amongst other explanations spatial planning remained steadfast as a non-statutory, area-related public sector task.

The 'Torremolinos Charter' adopted in 1983 by the European Conference of Ministers responsible for Regional Planning (CEMAT) presented one of the earliest definition of spatial planning. It states as follows:

1.2 Extended Definition of Terms

Regional/spatial planning gives geographical expression to the economic, social, cultural and ecological policies of society. It is at the same time a scientific discipline, an administrative technique and a policy developed as an interdisciplinary and comprehensive approach directed towards a balanced regional development and the physical organization of space according to an overall strategy.

The Royal Town Planning Institute offered two definitions. One is that 'spatial planning is critical thinking about space and place as the basis for action and intervention' and the other defines spatial planning as 'making of place and mediating of space' (Kunzmann 2004: 385). According to Wales Spatial Plan:

> 'Spatial planning is the consideration of what can and should happen where. It investigates the interaction of different policies and practices across regional space, and sets the role of places in a wider context. It goes well beyond "traditional" land use planning and sets out a strategic framework to guide future development and policy interventions whether or not these relate to formal land use control' (Welsh Assembly Government 2008: 3).

In an independent representation of spatial planning Adams et al. 2006 stated thus: 'spatial planning seems to reflect a more ambitious holistic approach to territorial development, incorporating all actors in a region to follow a joint vision for the development of a geographically-defined territory'. They also stated that 'spatial planning is a particular form of public policy, one that claims to be focused on the spatial dimensions of a wide range of other sectoral policies, from economic development, transportation and environmental protection through to health, culture and language'. Furthermore they explained that 'the holistic feature of spatial planning results from its cross-disciplinary nature, linking social and economic, as well as cultural and environmental dimensions of urban and regional development' (Adams et al. 2006: 50). Another independent representation states that Spatial Planning is the management of change, a political process by which a balance is sought between all interests involved, public and private, to resolve conflicting demands on space (Cilliers 2010: 73).

Given these definitions, it is upheld that spatial planning deals with space-activity relationships and the distribution of resource utilization. It is non-statutory and derives from spatial aspects of different categories of planning ranging from local land use plans to trans-national spatial development frameworks expressed in 'territorial planning' sometimes referred to as 'Territorial Cohesion'. Strategic spatial planning is identified to focus on the physical design of areas and the management of land use change through 'town planning schemes', which detailed the land uses allowed and their intensity.' (Todes et al. 2010: 416). Also comprehensive spatial planning is identified with enlarged content, which looks at the spatial dimension of all strategic policies and aim at integrating and coordinating all space-consuming activities in a geographic territory. It goes beyond traditional land use planning to bring together and integrate policies for the development and use of land with other policies and programs which influence the nature of places and how they function. Spatial planning hereafter refers to this definition (impression).

The inherent procedural/methodological renaissance and scoping define the point of departure for comprehensive spatial planning models that encouraged 'the design of spatial framework in which the evolution of society and economy could be accommodated (Albers 1986: 22)'. This elicits increased attention on understanding local contexts and formulating guiding frameworks that inform land development and management (Landman 2004: 163). However, with these transitions to policy analysis and emphasis on procedural changes, spatial planning is endangered as a tool for practical application in direct intervention in geographic space.

1.2.3 Space Economy

The space economy is contextualized from planning perspective. This effort took bearing from a traditional subfield of the discipline of geography referred to as economic geography. Ab initio economic geography is held to be the study of the location, distribution and spatial organization of economic activities across the world. Technically it refers to the mapping of economic activities in space. Already economic geography has taken a variety of approaches to many different subject matters including the core-periphery theory, the economics of urban form, the relationship between the environment and the economy and globalization.

Economic geography found expression in classical location theories of regional development. Within the 1930s–1950s periods Weber's (1929) location theory and spatial organization theory that highlights the work of Walter Christaller's (1933) central place theory, including growth theories represented in divergent and convergent hypothesis models especially Myrdal (1957) Cumulative causation theory and Perroux (1950) Growth pole theory, respectively, were notable approaches. The structuralist and institutionalist theories concerned with the input of politicians and planners in managing regional growth in their constituencies, which dominated the 1950s–1980s periods, had little to do with economic geography. Neoclassical theories, which rekindled attention on economic geography, evolved in the 1990s leading to Endogenous Growth Theories and Krugman's (1990) theory of New Economic Geography.

As economic inequalities are increasing spatially in the arising new economy, Krugman's theory introduced the spatial dimensions of regional growth and trade. It expressed concern for social, cultural and institutional factors in the spatial economy. The theory enabled economic geographers to study social and spatial divisions including the switch from manufacturing-based economies to the digital economy. They either engage in sophisticated spatial modelling or explain the apparently paradoxical emergence of industrial clusters in a contemporary context (Perrons 2004). Emphasis is on growth rather than development of regions in which case the actual impact of clusters on a region is given far less attention, relative to the focus on clustering of related activities in a region (Perrons 2004). Both approaches, however, acknowledge the importance of knowledge in the new economy, possible effects of externalities, and endogenous processes that generate increases in productivity. At the moment the development of economic geography drifts along those lines.

The space economy is further considered against the backdrop of urban ecology vis-à-vis urban economic functions. Similar to economic geography, urban ecology is concerned with the pattern of distribution of urban functions although it differs with considering distribution in space and goes further to address the dynamics in patterns of distribution given changes in critical variables such as planning controls. Thus unlike economic geography, the ecology of urban functions does not dwell on economic models, specifically descriptive input-output model, constructed to fit into the traditional models of the economy. Nourse (1968) indicated that this category of models, which isolate the critical variables causing change in urban economy, show tendencies. The point of departure lies with how the geographic distribution of economic activity will change with changes in the critical variables. This is why the ecology of urban economics is rather disposed to function with dynamic theories that predict resultant changes in patterns of distribution of urban economic functions within the urban spatial environment. Simulating this category of dynamic theories is outstanding and should be given priority attention in research agenda for planning, especially now that spatial imbalances that accrue from natural processes in the development of economic land use patterns are primarily responsible for theoretical predictions that now exist in urban economic analysis and model building.

The spatial expression of the space economy as an activity system draws from the definition of the urban environment as a system of relationships between three functional layers: the system of base which addresses the environmental factors and the culture and value system of the people in the locality; the deep structure, which addresses activity-space relationships; and, the superficial structure, which is the expression of the urban form in physical space (Okeke 2015). The three layers exist in a system of causal relationship and are therefore mutually dependent. Suffice it to say that the system of base feeds the deep structure and the deep structure in turn feeds the superficial structure. The superficial structure by way of reaction impacts the system of base sometimes adversely resulting in climate change, global warming, deforestation, etc.

As an activity system the space economy graphically relates to all three layers of the urban environment. At the system of base, it relates to the ecosystem of natural resource and their distribution in space. In planning studies, these resources are regarded as the economic base of the environment. Depending on the worldview and value system of the human capital, the use of the resource base reflects the livelihood support system. The support system determines space-activity relationships. At the realm of the deep structure of the environment, models of these relationships are constructed in economic space that exists in mental abstraction. The neoliberal economic models conceived in Africa within the 1970–2000 periods are typical examples. These models deal with space economy in abstract economic space as it is the case with Ann Roell Markusen's (1985) study approach. The models are normally translated into physical space through a planning process and this determines the superficial structure of the environment. Thus at the superficial level, the space economy finds expression in the distribution of people and activities as the society strategizes for economic productivity. This normally results in regionalization and the emergence of growth centres.

The space economy is not a static element and configuring it means getting a 'snapshot'. This mission is as difficult as aiming to hit a moving target. The fact remains that the three notions of space economy are mutually integrated. They subsist as natural ecosystems in which the abstract and practical elements are bound with the propulsive force released through planning rationality. This is why in the process of establishing a new order in the space economy major political and socio-economic transformations profoundly impact planning perspective. How it works is that any stimulus arising from major transformations automatically impacts the system of base. This kick-starts ripple effects that spread through the abstract to the practical realms of space economy. For instance, colonization and capitalism that were delivered in the context of imperialism impacted the worldview and value system of Africans at the outset and with the instrumentality of formal planning ultimately reworked the space economy at the practical realm. Neoliberalism as global economic orthodoxy introduced more stimulus with its crusade for liberalization and informality as operative standards in the system of base. As a result the space economy is further confounded.

1.2.4 Imperialism

There is no universally accepted theory of imperialism. Some even deny its existence and refuse to accept the broad definition of imperialism as political domination. Regardless, Hawksley (2004) classified the vast historiography of imperialism into four general schools each with differing explanations as to the operative concept and driving force behind the political restructuring of the world. The four schools identified are broadly economic, sociological, statist and cultural in character. Hawksley further indicates that each has a focus of inquiry that delineates what are core explanations of why imperialism occurs. Each makes its contribution to the debates on imperialism and political domination and all provide insights into the relationship of the governed to the governing. The knowledge of Hawksley's classification is enough. It is fruitless pursuing this knowledge further because the classification of imperialism with spatial character is imperative for this book.

Meanwhile the definition of imperialism is seldom given. Attention seems to focus on the instrumentality of imperialism and this has generated volumes of literature. The massive expansion of the area under the control of other powers has been the subject of much debate (Hawksley 2004). For purposes of examining the theoretical framework, timeframe seem to present the best line of argument and colonial experience coupled with the incidence of capitalist economy in the African region favours the twentieth century as the limit of retrospection.

1.2 Extended Definition of Terms

For most of the twentieth century, imperialism was seen to be a process that involved relations between developed states and areas of the world that were not yet states. In the early twentieth century imperialism is described as 'gunboat diplomacy' by Craven (2009). This expression captures efforts made to export capitalism into distant regions. Latter twentieth century imperialism assumed the outlook of social systems engineering. This relates to the destabilizing role of professional colonial officers who worked at the intermediate level within a colony or dependent territory and acted as part of the bureaucracy of an administrative district. Neo-imperialism was then conducted through the international regimes that coordinate policy and activity on behalf of the major international economic actors: the imperialism of capitalist interests acted to ensure the subjugation of the periphery, whatever the stated purpose of regimes (Hawksley 2004).

What is twenty-first century imperialism? According to Katz (2002) the notion of imperialism conceptualizes two types of problem: on the one hand, the relations of domination in operation between the capitalists of the Centre and the peoples of the periphery and on the other the links which prevail between the great imperialist powers at each stage of capitalism. Craven (2009) speculates that imperialism is the monopoly stage of capitalism. But Joseph Schumpeter (in Mommsen 1982: 23) feels capitalism is by nature anti-imperialist. Rather than being the cause of imperialism, capitalism was a direct contrast to it. (Hawksley 2004). Schumpeter argues that the territorial expansion associated with capitalism came from forces outside of the logic of capitalism but were supported by social and political factors outside of modern life. In their explanation of imperialism, the interface between Marxist and neo-Marxist theorists is the ambivalent role of the state. Dependency theorists in their contribution tend to connect imperialism with the neglected historical links of economic extraction between the states of the developed and the developing world.

According to the Merriam-Webster dictionary, imperialism is defined as, 'The policy, practice, or advocacy of extending the power and dominion of a nation especially by direct territorial acquisitions or by gaining indirect control over the political or economic life of other areas', This neocolonial vision is today embedded in global mercantilism and operationalized through trade liberalization. Schumpeter (in Mommsen 1982: 22) defined imperialism as 'the objectless disposition of the part of the state to unlimited forcible expansion'. Hawksley (2004) extends this definition with the expression 'that imperial expansion was an irrational action inclined toward war and conquest, and due mainly to the survival of political structures from the era of absolutism'. Somehow, imperialism is associated with social rewards, state building and nationalism, but above all a sort of pre-modern desire for conquest (Hawksley 2004).

Without forgetting the conditional and relative value of all definitions in general, which can never embrace all the concatenations of a phenomenon in its full development, Craven (2009) gave a definition of imperialism that will include the following five of its basic features:

i. the concentration of production and capital has developed to such a high stage that it has created monopolies which play a decisive role in economic life;
ii. the merging of bank capital with industrial capital, and the creation, on the basis of this 'finance capital', of a financial oligarchy;
iii. the export of capital as distinguished from the export of commodities acquires exceptional importance;
iv. the formation of international monopolist capitalist associations which share the world among themselves and
v. the territorial division of the whole world among the biggest capitalist powers is completed.

Craven (2009) went ahead to define imperialism as a system and not simply a set of policies. Its ultimate goal is to ensure that the former imperial states still maintain economic dominance over their former colonies (Devon 2011). This holds implications on how imperialism is practiced.

Hawksley (2004) provides an evolution of how imperialism is practiced based on four criteria: the treatment of people under the rule of the colonizers; the level of physical coercion present; the sophistication of systems of government; and, the stated intentions of the imperial power. Consecutively, guided by the criteria, he first identified imperialism of plunder which occupies territory through force of arms, has scant regard for the peoples that it dominates and frequently enslaves them, works them to death or kills them when taking land. The sole aim of the imperialism of plunder in acquiring territory is to exploit the riches it possesses. Second is the imperialism of private commerce. It requires both the state and private companies and consists of a trading base, a fort or a 'factory', often with a small defensive perimeter. The purpose of imperialism by private commerce is profit. The aims of imperialism by private commerce are to run a successful business and to create profit for directors, shareholders, and for the officers of the company, the workers and governors who run the day-to-day business.

Expansionist imperialism which is primarily concerned with territorial acquisition is the third. The aims of Expansionist imperialism are to acquire territory and to create new markets. This model of imperialism is reminiscent of the activities of the British, French and Germans in Africa in the historical period known as the 'New Imperialism' (1870–1914). Fourth is the Imperialism of administration. The imperialism of administration moves away from the use of physical force in favour of increasingly sophisticated techniques of social management. Structured coercion of governmentality is deployed. The main motive is not so much the creation of profit for the colonial state but the linking of the region to the international economy. Imperialism as administration demands nothing less than the wholesale transformation of society to reflect the priorities of the colonizers. The last is imperialism as international administration. In this case multinational organizations, particularly the United Nation (UN), are called upon to be an administering authority. Hitherto the UN is content to monitor states in exercising their responsibilities as trustees of other territories (Hawksley 2004).

Imperialism reflects in the use of space. This brings the planning perspective to bear in the consideration of the theoretical framework of imperialism. Incidentally it is yet to be established in planning literature that any scholarly work has been done in this direction. Perhaps this is because planning discuss tend to focus or indeed tend to be limited to the scientific realm. It rarely extends to the meta-theoretical level which demands the extension of the theoretical arsenal of planning scholarship beyond the theories of planning. The working definition of imperialism from planning perspective in this book perceives imperialism as a system of spatial expression of dependence. Its basic features point to spatial dualities and disconnects in the space economy as well as the alignment of development corridors and urban form that responds to the expansionist program of market-space economy for Euro-America and Chinese mercantilism.

1.2.5 Neoliberalism

Neoliberalism represents a re-assertion of the fundamental beliefs of the liberal political economy that was the dominant political ideology of the nineteenth century, above all in Britain and the United States (Clark 2013: 7).The term 'neo-liberalism' was originally coined in 1938 by the German scholar Alexander Rüstow at the Colloque Walter Lippmann. It was used in the colloquium to represent 'the priority of the price mechanism, the free enterprise, the system of competition and a strong and impartial state'. The new nature of neoliberalism presents it either as a philosophy or ideology or as an attitude or as a development model or as an academic paradigm, suggesting that it can assume extreme forms including cyber-liberalism which overlaps with semi-religious beliefs in the interconnectedness of the cosmos. Ultimately it applies market metaphor in the perception of the world.

Neoliberalism is an attitude rather than an economic reality. The term is used in several senses: as a development model it refers to the rejection of structuralist economics in favour of the Washington Consensus; as an ideology the term is used to denote a conception of freedom as an overarching social value associated with reducing state functions to those of a minimal state; and finally as an academic paradigm the term is closely related to neoclassical economic theory (Taylor and Jordan 2009). The general ethical precept of neoliberalism can be summarized approximately as follows:

- 'act in conformity with market forces'
- 'within this limit, act also to maximize the opportunity for others to conform to the market forces generated by your action'
- 'hold no other goals'

Martinez and García (2000) identified the principles of neoliberalism to include: the rule of the market, cutting public expenditure for social services, deregulation, privatization and eliminating the concept of 'the public good' or 'community'. Neoliberalism assumes freedom of access to law and information. This is contentious because it has associated costs in which the rich has advantage, resulting in social stratification also known as class. The tendency to create and strengthen class has resulted in some claiming that neoliberalism is a class project, designed to impose class on society through liberalism.

In the past two decades, according to the Boas and Gans-Morse study of 148 journal articles, neoliberalism is almost never defined but used in several senses to describe ideology, economic theory, development theory, or economic reform policy. Neoliberalism is certainly a form of free-market neoclassical economic theory, but it is quite difficult to pin down further than that, especially since neoliberal governments and economists carefully avoided referring to themselves as neoliberals (Mohammadzadeh 2011). Further definitions indicate that neoliberalism is essentially a philosophy and an economic concept used by political entities to determine corporate relationships in regional and global development processes. As Milton Friedman put it, neoliberalism rests on the 'elementary proposition that both parties to an economic transaction benefit from it, provided the transaction is bilaterally voluntary and informed' (Friedman 1962: 55).

With regards to the goal of neoliberalism, there are indications in literature that neoliberalism is a form of neocolonization that has nothing to do with liberalism (see David 2005). This view leans on the consistency of worsening conditions in countries where 'neo-liberal' policies are imposed and on tendencies where more developed countries can exploit the less developed countries. It is noticed however that this perception of neoliberalism is not a consensus because even opponents do not agree. They present counter interpretation of the performance of neoliberalism against the interests of the country being 'saved'. It is argued that local elites could take advantage of neoliberal reforms. In the same vein neoliberal fundamentalists could collude unwittingly with multinational corporations to exploit political economies in the global south. This is erroneously blamed on neo-imperial plot. Nevertheless neoliberalism cannot be fully exonerated from neo-imperial plot in so far as developing countries are disadvantaged and vulnerable, dependent and marginalized in the world-system.

1.3 Methodology

This work was drawn primarily from the research conducted within 2010–2015 period at North West University, Potchefstroom campus, South Africa. It adopts world-systems analytical approach to engage qualitative and quantitative research. A five-stage research activity system is envisaged. It includes the following:

i. Postulating African renaissance as mission for cities in Africa
ii. Rethinking neoliberalism as development ideology (and meta-theory for planning)

iii. Postulating neo-mercantilism as development ideology for Africa (and meta-theory for planning)
iv. Introducing the **neo-mercantile planning theory**
v. New knowledge

The five-stage research activities are mutually dependent. The theoretical framework will be supported with graphic illustrations including tables, figures, graphs, charts, etc.

1.4 Analytical Approach

The analytical approach in this work is reminiscent of the world-system analysis used in social sciences to examine historical systems. World-system analysis is a holistic approach that seeks to explain the dynamics of the 'capitalist world-economy' as a 'total social system'. It originated in the 1970s around the work of Immanuel Wallerstein. Wallerstein (2004) indicates that the core concepts of world-systems analysis were not new but borrowed from different disciplines, such as history (the term world-system itself), economics (unequal exchange), sociology (classes) and politics (balance of powers). He further explained that what was new was the articulation of these concepts into a coherent holistic perspective. The integrative process basically changed the unit of analysis from political structures (i.e. nation-states) to marketization and privatization, incorporated core-periphery concepts and encouraged multidimensional and multi-secular (i.e. historical) analysis. These attributes underpin the overarching global influence on paradigm shifts in planning.

There is no gainsaying that global influences are largely responsible for the new perspective in planning that seeks to reinvent planning in line with neoliberal values. Yet at the moment there is no indication that these influences inform current perspectives of analyzing trends in spatial planning in Africa, perhaps the most threatened region in the world system. Analytical procedures are preoccupied with addressing local influences centred on sourcing plan implementation, while pre-dating global influences hijack the system to push its agenda. Thus, the delivery of docile planning as required in the world system is underway, strongly guided by imperial systems deployed in historical systems—global capitalism—to suppress dissenting views.

World-system analysis is focused on global influence, therefore it is best suited to analyze current trends in development perspectives. It is intended that this approach be used to analyze the evolution in urban planning paradigms in Africa. It is holistic and executed in the context of epistemological foundations of imperialism found in historical systems of African civilization. This analytical orientation assumes an historical perspective of examining paradigm shifts following major socio-economic transformations at global and local levels. The global analysis is complemented by perception analysis using the MCA approach and own assessment to assess the context and content of planning initiatives in selected African countries. Meanwhile all analytical procedures were guided by the mindset of instituting spatial regional integration in Africa.

1.5 Synthesis

The set of findings generated in the study served as theoretical foundation for the proposed spatial development paradigm. Through a visioning process the applicability of the proposed paradigm is tested in the conceptualization of a

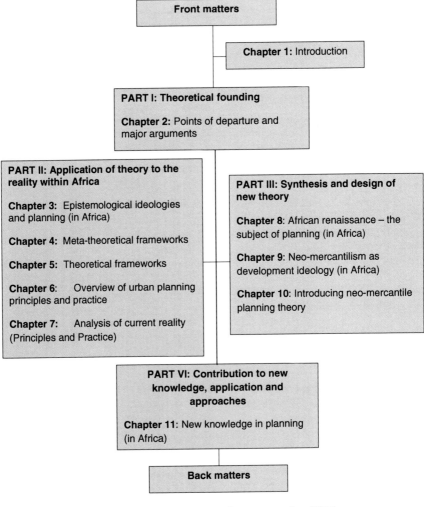

Fig. 1.3 The structure of the book report. *Source* Own construction (2016)

1.5 Synthesis

hypothetical planning paradigm for Africa codenamed the 'neo-African paradigm'. The initiative, perspective and framework instruments derived from neo-African paradigm wholly represent original new knowledge which this study contributes to initiate the development of planning theory for Africa.

1.6 Structure of Book Report

The book is presented in 6 sections and 9 chapters. A schematic illustration of the structure of the book report is shown in Fig. 1.3 below.

1.7 Conclusion

The need to reinforce the theoretical arsenal of planning scholarship has been severally diagnosed. The practice of relying on theoretical perspectives developed by economists, sociologists and political scientists to conceptualize planning initiatives is called to serious question. The earlier planning scholarship cuts across practical, theoretical and meta-theoretical levels the sooner the orientation of planning adopts a professional outlook. At the moment the consideration of the practical and theoretical levels is the vogue. The exercise of planning at these levels submits to the external interpretation of its meta-theory. This book crusades the reversal of this trend. Planners should engage the meta-theory of planning directly. The meta-theory of planning essentially refers to development ideologies. Planning is normally the delivery tool of these ideologies. In some normal instances the ontological responsibility of planning is dialectically opposed to the general ethical precepts of development ideologies, except perhaps the planning system is compromised or malfunctions.

This book pioneers the inclusion of the meta-theory of planning in the analysis of spatial planning in Africa. Precisely, appraising neoliberalism as meta-theory of planning is the central element of investigation. This methodology of investigation is deployed to determine the platform for assessing current planning actions in Africa. Viewed locally and within the limits of scientific consideration, current actions earn some measure of justification. However, given the realities of global political economy, gleaned from the meta-theory of planning, current actions are regarded with some reservations. It is too survivalist and framed to attend to the externalities of neoliberal economy in Africa. Its capacity to bring Africa out of poverty and dependence is called to question. These issues impact on the impression of adequacy associated with current actions and raises arguments that moderate their relevance. Thus, current actions are ideally interim measures for application at local levels while robust planning instruments are theorized to engage planning requirement for re-engineering the space economy, to make them resilient to imperialism. This addresses the aspiration of redressing poverty, which the current and new actions share. But while current action is palliative the new direction is curative. The new direction contends with the integration of supporting spatial system within nation-states, in anticipation of Africa's productive participation in new regionalism.

This book introduces the neo-mercantile planning theory. The theory is the product of an analysis of spatial development paradigm for enhancing regional integration within national and it is supporting spatial systems in Africa. The study was conducted within 2010–2015 period at North West University, Potchefstroom campus, South Africa. It was motivated by the need to enhance urban productivity in Africa. The major challenge was to establish that formal planning has not outlived its usefulness in Africa, contrary to the impression created by neoliberal content drivers. This effort pre-empted the apprehension of the intellectual community towards paradigm shift that popularized participatory process in planning. The planning initiatives built on participatory framework were found to lack the capacity to deliver integration. Against this backdrop, the resilience of formal master planning grew in planning literature.

The provisions made for the implementation of the theory requires the political will of the committee of nations at the African Union (AU). It is anticipated that the theory will guide the preparation of African Spatial Development Perspective (ASDP). The AU needs the ASDP to coordinate its functions as the administrative authority for regional integration in Africa.

The methodology engaged qualitative and quantitative research.

References

Adams A, Alden J, Harris N (2006) Regional development and spatial planning in an enlarged European Union. Ashgate, Hampshire, 8, 13, 44, 50p
Albers G (1986) Changes in German town planning: a review of the last sixty years. Town Plann Rev 57(1):17–34
Altschuler AA (1965) The city planning process: a political analysis. Cornell University Press, Ithaca, NY
Arbury J (2005) From urban sprawl to compact city—an analysis of urban growth management in Auckland. Thesis, PhD Auckland University, Auckland
Arpke F (1942) Land use control in the urban fringe of Portland, Oregon. J Land Public Utility Econ 18(4):468–480
http://assets.wwf.org.uk/downloads/ma_msp_wa.pdf Date of access 23rd May 2014
Bairoch P (1975) The economic development of the third world since 1900. Methuen, London
Berke PR (2002) Does sustainable development offer a new direction for planning? Challenges for the twenty-first century. J Plann Lit 17(1):22–24, 29
Christaller W (1933) Central places in Southern Germany' translated by CW Baskin. Prentice-Hall, Englewood Cliffs, New Jersey
Chukuezi CO (2010) Urban informal sector and unemployment in third world cities: the situation in Nigeria. Asian Social Science. 6(8):133
Cilliers EJ (2010) Creating sustainable urban form: the urban development boundary as a planning tool for sustainable urban form: implications for the Gauteng City Region. Vdm Verlag, South Africa

References

Clark S (2013) The ideological foundations of neo-liberalism. Available at: www.heathwoodpress.com/ Date of access 2 Jan 2016

Craven JM (2009) Neoclassical economics and neoliberalism as neo-imperialism. A presentation to Institute of Marxism Research of the Chinese Academy of Social Sciences. Available at https://jimcraven10.wordpress.com/. Date of access 2 Jan 2016

David H (2005) A brief history of neo-liberalism. Oxford University Press, Oxford, 235p

Dembele DM (1998) Africa in the twenty-first century. 9th General Assembly CODESRIA Bulletin ISSN 0850-8712 No 1, pp 10–14

Devon DB (2011) Neo-colonialism, imperialism and resistance in the 21st century. global research. Available at www.globalresearch.ca/. Date of access 2 Jan 2016

Edwardo EE (1990) Community design and the culture of cities. Cambridge University press, Cambridge, 356p

Frank AG (1971) Capitalism and underdevelopment in Latin America. Penguin, Harmondworts

Friedman M (1962) Capitalism and freedom. University of Chicago Press, Chicago

Friedmann J (1965) Response to Altshuler: comprehensive planning, as a process. J Am Institute Planners 31(3):195–197

Graves PE, Sexton RL (1979) Overurbanization and its relation to economic growth for less developed countries. Econ Forum 8(1):95–100 (Reprinted in Ghosh KP 1984:160–166)

Hansen WB (1968) Metropolitan planning and the new comprehensiveness. J Am Institute Planners 34(5):295–302

Hawksley C (2004) Conceptualizing imperialism in the 21st century. Being a paper presented at the 2004 Australian Political Studies Association Conference at the University of Adelaide. Available at https://www.adelaide.edu.au/apsa/docs_papers/Others/Hawksley.pdf. Date of access 26 Nov 2015

Healey P (1997) The revival of strategic spatial planning in Europe. In: Healey P, Khakee A, Motte A, Needham B (eds) Making strategic spatial plans: innovation in Europe. UCL Press, London, pp. 3–19

Hicks J (1998) Enhancing the productivity of urban Africa. In: Proceedings of an International Conference Research Community for the Habitat Agenda. Linking Research and Policy for the Sustainability of Human Settlement held in Geneva, July, 6–8

Katz C (2002) Imperialism in the 21st century. Available at www.katz.lahaine.org/. Date of access 2 Jan 2016

Krugman P (1990) The role of geography in development. Int Reg Sci Rev 22(2):142–161

Kunzmann KR (2004) Culture, creativity, and spatial planning. Town Plann Rev 75(4):384–385, 397p

Landman K (2004) Gated communities in South Africa: the challenge for spatial planning and land use management. Town Plann Rev 75(2):163

Lowder S (1986) Inside third world cities. Chrome Helm, London

Majekodunmi A, Adejuwon KD (2012) Globalization and African political economy: the Nigerian experience. Int J Acad Res Bus Soc Sci 2(8):189–206

Markusen A (1985) Profit cycles, oligopoly, and regional development. MIT Press, Cambridge, MA

Martinez E, Garcia A (2000) What is 'Neo-liberalism?'—a brief definition. Available at: http://www.globalexchange.org/resources/econ101/neoliberalismdefined. Date of access 18 Aug 2013

Meyerson M (1973) Building the middle-range bridge for comprehensive planning. In: Faludi A (ed) A Reader in planning theory. Pergamum, Oxford and New York

Mohammadzadeh M (2011) Neo-liberal planning in practice: urban development with/without a plan (A post-structural investigation of Dubai's development). Paper Presented in Track 9 (Spatial Policies and Land Use Planning) at the 3rd World Planning Schools Congress, Perth (WA), 4–8 July 2011

Mommsen WJ (1982) Theories of Imperialism. Weidenfeld and Nicolson, London

Myrdal G (1957) Economic theory and underdeveloped regions. Duckworth, London

Nabudere DW (2003) Towards a new model of production—an alternative to NEPAD. 14th Biennial Congress of AAPS. Durban, South Africa. www.mpai.ac.ug. Date of access 8 July 2011

Nourse HO (1968) Regional economics. McGraw-Hill, New York

Okeke DC (2015) An analysis of spatial development paradigm for enhancing regional integration within national and its supporting spatial systems in Africa. Thesis submitted in fulfilment of the requirements for the degree *Philosophiae Doctor* in Urban and Regional Planning at the Potchefstroom Campus of the North-West University, South Africa

Owusu F (2007) Conceptualizing livelihood strategies in African cities: planning and development implications of multiple livelihood strategies. J Plann Edu Res 26(4):450–463

Perrons D (2004) Understanding social and spatial divisions in the new economy: new media clusters and the digital divide. Econ Geogr 80(1):45–61

Perroux F (1950) Economic space: theory and applications. Quart J Econ 64(1):89–104

Roberts B (1978) Of peasants. Edward Arnold, London

Roy A (2009) Why India cannot plan its cities: informality, insurgence and the idiom of urbanization. Plann Theory 8(1):76–87

Santos N (1979) The shared space. Methuen, London

Simone A (1998) Urban processes and change in Africa. CODESRIA Working Papers 3/97 Senegal, 12p

Taylor CB, Jordan G (2009) Neoliberalism: from new liberal philosophy to anti-liberal slogan. Stud Comp Int Dev 44(2):144

Todes A, Karam A, Klug N, Malaza N (2010) Beyond master planning? New approaches to spatial planning in Ekurhuleni, South Africa. Habitat Int 34:414–416, 419p

Wallerstein I (2004) World-systems analysis: an introduction. Duke University Press, USA, 128p

Wallerstein I (1974) The modern world system, vol I. Academic Press, New York

Watson V (2009) 'The planned city sweeps the poor away…': urban planning and 21st century urbanization. Progress Plann 72:153–193

Weber A (1929) Theory of the location of industries. University of Chicago Press, Chicago

Welsh Assembly Government 2008 Peoples, Places Furture - Wales Spatial Plan. Available at www.torfaen-gov.uk/....Planning/SD78 Date of access 20th July 2016.

World Bank (1989) Sub-Saharan Africa: from crisis to sustainable growth. The World Bank, Washington, D.C

Part I
Theoretical Founding

Chapter 2
Points of Departure and Major Arguments

Abstract The epistemologies of imperial mindset, imperial space economy and imperial cities are critical points of departure in determining the way forward for planning in Africa. The revision of these epistemologies, entrenched under the growing influence of informality, is assumed to be the challenge ahead of planning intervention in Africa. In search of an appropriate planning perspective, this work assumes that form-based planning attributes are not significantly resilient in planning within spatial systems in Africa. Although contrary to current trend, the work argues that in so far as a new spatial planning perspective is without the form element it lacks merit to initiate a theoretical evolution in spatial planning. As such it further argues that formal expertise knowledge should take precedence over informal expertise knowledge in planning. These dispositions point to the need to revisit neoliberalism as development ideology and thinking instrument for planning in Africa.

Keywords Epistemologies · Imperial · Formal planning · Neoliberalism · Spatial planning · Institutionalist · Africa

2.1 Points of Departure

The points of departure identified for this work are rooted in the epistemological foundations of imperialism found in Africa. These epistemologies are scattered in four of the six historical systems discerned in African civilization. The historical systems are: the period prior to tenth century when medieval African Kingdoms flourished; between tenth century and fifteenth century, the mercantilist period marked by the Trans-Sahara Trade; between fifteenth century and mid-nineteenth century, the slave trade period; between mid-nineteenth century and 1950, the capitalist colonial period; between 1950s and 1980s, the independence decade and partial Keynesian period; and from 1980s until now, the neoliberal period. The mercantilist period marked the watershed for imperialism in Africa. As imperialism transited along the historic systems it built support structures that are manifest in spatial systems and planning approaches.

By mid-fifteenth century in the mercantilist period the notion of imperialism as private commerce was sworn following contact with European merchants. Imperialism of commerce consists of a trading base, a fort or a 'factory', often with a small defensive perimeter (Hawksley 2004: 18). The purpose of imperialism by private commerce is profit. The Europeans in their trade relations sort for control and trade monopolies. This followed strategic change of development ideology to classical liberalism, which redefined trade relations in favour of European mercantilism. The change caused fundamental changes in the system of base of traditional African cities. The first phase of imperial space economy developed by accretion within this period.

The capitalist colonial period, which ensued mid-nineteenth century, witnessed the change to material capitalism in which neo-mercantilism served as trade strategy to consolidate European mercantilism. Imperialism as commerce remained. The colonial system restructured peasant agriculture, introduced new administrative systems, and changed the pattern of urbanization with the incidence of imperial colonial towns (Rakodi 1997). In the process, the integrated cosmology of traditional Africa was replaced with single-minded utilitarian objectives which produced utilitarian designs for cities in Africa. The design options bulldozed cultural symbols, behaviour and beliefs that determined the base of traditional African cities. Cities in Africa became hybrids, an inevitable product of intervening culture and policy formulation hegemony from abroad. These events marked the second phase in the development of imperial space economy. At this time formal planning was introduced to deliver colonial interest in the use of space and facilitate the layout of infrastructure to enhance extraction and exploitative trade relations.

The independence decade (1950–1960) within the immediate post-colonial period witnessed transition in the manifestation of imperialism. Imperialism drops physical coercion in favour of social management to deliver structured coercion of governmentality. Imperialism of administration interprets this system in which the emerging ruling class at independence continued acting the script already written by the departing colonizers (Majekodunmi and Adejuwon 2012: 197). The changes in the traditional system of base vis-à-vis culture, value system and worldview remained valid and resilient to reversion. Shortly, after the independence decade the approach of neoliberalism as development ideology and economic orthodoxy engineered the criticisms of formal master planning to pave way for participatory planning. Participatory planning is all about democratizing planning decisions under the guise of enhancing plan implementation when indeed it facilitates exploitative partnerships. It is epitomized in neoliberal planning theory.

The era of neoliberalism and neoliberal planning theory crystalized in the 1980s and since then it has endeavoured precariously to guide development. So far the main points of departure in this period are the incidence of informality in the context of imperialism as international administration. The doctrine of informality impacts the economy, planning and spatial systems. Thus, informal economic sector, informality in planning and the informalization of cities gained momentum.

The spatial distortion this situation creates in the economic landscape of the urban regions is unprecedented.

The epistemologies of imperial mindset, imperial space economy and imperial cities are all critical points of departure that demand attention. This is because the resultant dominantly introverted urbanization and the informal and extroverted urban economy seriously challenge growth and productivity in African countries. The peculiar pattern of growth indicates positive GDP growth amidst stagnant or negative productivity measured in declining per capita GDP, growing poverty, debilitating unemployment and high Gini coefficients. This is typical in Nigeria it and South Africa and perhaps to a lesser extent in Tanzania and perhaps Egypt.

2.2 Assumptions

Spatial planning instruments that contend for relevance in redressing distortions in the structure of urban regions and reverse the extroversion identified in the development of space economy in Africa are arranged into two categories, namely form-based (formal) and non-form-based (pragmatic) planning instruments.

For purposes of clarity, form-based instruments operate with the principles of form and function in planning for land use interventions. Hence upholds planning rationality as determinant for integrated development of the urban region. On the other hand, non-form-based planning instruments dwell on informality, which disregards planning rationality in principle. It upholds market forces as a determinant factor for land use intervention. The two schools of thought represent alternative approaches to spatial planning for integrated development of space economy. Both approaches pursue economic growth but it is argued that the former does so in the context of shaping the urban region in the spatio-physical sense for nation-building, unlike the latter which does so strictly in the economic sense for private profitability.

Hitherto the space economy—that is the development of the urban region—in Africa has been fraught with sprawl and disconnects responsible for urban productivity decline and more so in the context of a dependent capitalism underpinned by the consumer economy. The revision of these attributes, entrenched under the growing influence of informality, is assumed to be the challenge ahead of planning intervention in Africa. Essentially, planning interventions should rework the space economy and make it compatible with introverted economic growth intended to relieve Africa from dependent capitalism. Delivering on this milestone(s) is assumed to be the primary function of the planning approach being targeted. In other words, the planning approach, in practical terms, will deliver an integrated development of the urban region in spatio-physical terms for enhanced productivity in Africa.

This work assumes that form-based planning attributes are not significantly resilient in planning within spatial systems in Africa.

2.3 Major Arguments

The African development surface is known to manifest distortions that are responsible for the urban productivity decline in the region. Indeed, these distortions, according to Hicks (1998) leave the legacy of isolated urban hierarchies with limited linkages in the urban region mainly in the form of urban-rural dichotomy and fragmentation of the private sector 'with extroverted modern sector' sparsely 'related with the local economy'. Therefore, the global objective of development action for Africa is invariably focused on integrated regional development. Viewed from the planning perspective this objective focuses on spatial integration underpinned by territorial planning.

The global objective of integrated regional development is connected with sourcing enhanced productivity through the introversion of the economies of urban Africa. Succinctly put, the objective indicates an African renaissance. Efforts towards realizing the global objective have led to new partnerships in African development (NEPAD) initiatives with its political and economic reforms. This provides a compelling opportunity to consider appropriate spatial paradigms that will translate the global objective into space. The visionary process required to theorize the paradigm for the spatial planning intervention is seen to be in dilemma due to paradigm shifts in planning linked with the incidence of neoliberal planning.

In spite of the predating status of neoliberal planning perspective, it is clear from literature that the master planning paradigm remains resilient. In practical terms, process-oriented planning mainstreamed in the visioning process for NEPAD implementation, is upheld against formal planning perspectives upheld in national development planning. The resilience of formal planning paradigm preoccupies this work. Thus, applying this knowledge base to determine an appropriate planning paradigm in which participation is mainstreamed for the delivery of spatial regional integration within spatial systems in Africa represents the core problem of this work.

2.3.1 Theoretical Perspective

The cyclical evolution noticed in the development of planning theory from classic–rational–neo-classic, which is in tandem with the evolution from pre-modern, modern and post-modern periods in planning epistemology, is not a coincidence. Equally illuminating is the synchronizing evolution of urban design in planning represented by the changing orientation of urbanism from old urbanism—traditional urbanism—new urbanism. All three categories of evolutions are driven by new facts generated either through environmental determinism or humanistic interventionist activities or epistemologies of imperialism. The combination of neoclassical planning theory and new-urbanism acting in the post-modern period in planning provides a lead to the contemporary emphasis on the use of space in spatial planning. Given their provisions the essence of planning without interferences from new perspectives such as neoliberal planning remains within the realm of morphology, hence the focus on the city.

Spatial planning ontologically is form-based and the integrity of planning rationality rests on this premise without prejudice to pluralism which is the hallmark of new perspectives in planning. Spatial planning remains an art and a science with the explicit aim to manage the use of space. This work argues that in so far as a new spatial planning perspective is without the form element, it lacks merit to initiate a theoretical evolution in spatial planning. Otherwise sustainable urbanism would have lost impetus under pressure from neoliberal planning. The hard reality of the limited role of new perspectives in driving evolution in spatial planning dawned on the use of neoliberal planning in Africa. The new planning perspectives do not have the integrative capacity to deal with the creative planning requirement of sustainable urban development. However, an interface could be sought that does not usurp the principles of form and function in spatial planning and this is where the African region misses the mark, unlike counter-part regions in the global north. In all of their commitment to neoliberal planning it is understood in these regions that form and function cannot be compromised hence the complementary role of urbanism.

It is noteworthy at this juncture that a participatory process is not necessarily what makes neoliberal planning a new perspective in planning. It is indeed a change in value systems associated with liberalization in global economy and planning outlook that is increasingly project-oriented and existential and the commitment to investigate development in a deregulated spatial planning context which identifies it as a new perspective. Participatory process plays a facilitating role and this perhaps explains the tango of neoliberal planning with the substrate of informality. On second thoughts, neoliberal planning is all about access to the control of space economy.

Informality is somehow connected with new perspectives in planning, hence the expression *informal planning*. Theoretically speaking, informal planning is an aberration amounting to parallel planning systems, although it could seek legitimacy on account of planning practised during the popular design tradition in the earlier segment of the pre-modernist planning period. Indeed, spatial planning of the earlier epoch, although practised in informal circumstances, was nevertheless orthodox spatial planning. The present dispensation argues that informality could be accommodated; however, as hypothetical design simulation(s). The simulations will be based on educated assumptions led by formal expertise/knowledge in planning and used as planning instrument for enhancing participation at the inception of spatial planning intervention. In this way planning principles are not compromised yet participation is not impaired.

Also, making growth in the context of territorial development theoretically compelling for spatial planning should inform new perspectives in spatial planning theory for the African region. Such developmental or indeed applied planning theory will drive the appropriate spatial paradigm to redress the distortions in the peculiar context of Africa's development surface.

2.3.2 Analytical Perspective

The quiet revolution in planning theory identified in the 1980s revolved around the determination of participatory process in planning. The primary argument leaned on poor plan implementation as rationale for change. At the turn of the twenty-first century, the progression of the argument lead to the consideration of informality in planning. This was short-lived thus paving way for the gradual shift of emphasis to conceptual issues such as spatial justice and the resilient cities concepts. The latest entry is new regionalism. The problem with the arguments behind these concepts is that it is not clear if the concepts do not bear neo-imperial plot as hidden agenda.

Regarding poor plan implementation, the problem is instead related to sharp practices in funding mechanisms the circumvention of which indeed elicits the need for participation. *Ab initio* participation is rooted in a breakdown of trust in the manipulative planning system that is commonplace in Africa and which in technical terms is participatory in nature. Also participatory process is known to facilitate imperialism as international administration. It renders planning vulnerable to market force and in the process allows funding mechanisms to play a systemic role in plan preparation and implementation. This explains the preference for project planning.

Plan preparation which is becoming increasingly desktop is bedevilled by mediocrity and complicated by quackery and charlatanism. These vices in planning highlight the situation when untrained personnel engage in taking planning decisions. The interactive participation advocated in neoliberal planning tends to provide legitimization for this syndrome. Continued liberalization of planning decisions is likely to be counter-productive in modelling the urban region and shaping the city as it is the case with the application of IDP/SDF in South Africa. These design requirements are technical issues that demand a lot more than political analysis. Regardless the argument that the availability of reliable formal expertise knowledge could be limited in some African countries such as the DRC, Angola, Ethiopia, etc., this work argues that formal expertise knowledge should take precedence over informal expertise knowledge in planning. Informal expertise/knowledge should play a facilitating role in principle, on the basis of which an appropriate interface will be found for political and technical analysis in planning.

The argument in favour of classic participatory formal planning is mindful of the repressive attitude towards it and the high nuisance value of this attitude, which is responsible for undermining formal master planning. The argument builds on the momentum of the new master planning to uphold visionary planning in preference to the current vogue of neoliberal existential planning.

2.4 Conclusion

Given these epistemological points of departure, the choice of the best line of action presents serious challenges because significant structural changes are required to rework the status quo. In the first instance, there is a need to revisit neoliberalism as development ideology and thinking instrument for planning in Africa. Also critical is the need to condition the mission of cities to address the global objective of African renaissance. Priority action lies in conditioning the theoretical framework for spatial planning in Africa and synchronize it with the fundamentals of new formal planning, which mainstreams participation. It is against this backdrop that a spatial development paradigm for Africa will be formulated. The planning paradigm formulation presumably will follow a process that is based on new facts driven either by environmental determinism or by humanistic interventionist activities. In essence, the work is not chasing a new perspective rather in the light of new facts it is theorizing an alternative paradigm that will deliver African renaissance.

References

Hawksley C (2004) Conceptualizing Imperialism in the 21st century. Being a paper presented at the 2004 Australian Political Studies Association Conference at the University of Adelaide. Available at: https://www.adelaide.edu.au/apsa/docs_papers/Others/Hawksley.pdf. Date of access 26 Nov 2015

Hicks J (1998) Enhancing the productivity of urban Africa. In: Proceedings of an international conference research community for the habitat agenda. Linking research and policy for the sustainability of human settlement held in Geneva, July, 6–8

Majekodunmi A, Adejuwon KD (2012) Globalization and African political economy: the Nigerian experience. Int J Acad Res Bus Soc Sci 2(8):189–206

Rakodi C (1997) 2 Global forces, urban change, and urban management in Africa. In: Carole R (ed) The urban challenge in Africa: growth and management of its large cities. United Nations University Press, Tokyo

Part II
Application of Theory to the Reality Within Africa

Chapter 3
Epistemological Ideologies and Planning (in Africa)

Abstract Epistemological foundations and ideologies of imperial space economy and informality are the bane of planning in Africa. These epistemologies draw from development ideologies, which reside in the meta-theoretical realm. The disregard of the meta-theory of planning is responsible for the gross misreading of planning. This explains the neglect of urban planning in favour of investment planning and environmental management. The urban environment ceases to be the subject of planning and the object of planning transits from spatial integration to poverty alleviation. Although the current informalization of cities has no theoretical basis for growth, the planning scene continues to tend towards neoliberal participatory planning for project development in environmental management. This widens the lacuna in urban planning and deepens urban crisis in Africa.

Keywords Epistemological · Informalization · Meta-theoretical · Diaspora · Traditional · Informality

3.1 Imperial Space Economy

Prior to contact with western civilization in the mercantile period marked by trade with Arab merchants, the structure of the space economy was determined by the spatial segregation of homeland territories (comprising resource and production areas) and commercial centres located at the interface of adjourning homeland territories and linked with regional roads. The pattern of homeland settlements is reminiscent of the city state concept which serves as a centre of production with civic identity and territorial boundaries under a charismatic leadership that controlled the affairs of state. The commercial centres, which serve as trading outposts, were impermanent confluence centres located at neutral grounds outside the boundaries of sovereign entities (homeland). These centres that are occupied by transient population lacked civic identity and by normative standards not qualified as an enclave (homeland) settlement. The enclave (homeland) city states were the primary settlements, the focus of production activities and not the centres of commerce. This structure was borne out of the normative initiative in space economy that cradled African civilization.

The imperial control of space economy ensued during slave trade period in mid-fifteenth century at the hills of Portuguese mercantilism in mid-fifteenth century and followed by British and French traders in sixteenth century. The Europeans in their capitalist trade relations sort for control and trade monopolies. It is reported that they played one kingdom off against another to prevent the emergence of powerful state (Guglar and Flanagan 1978). Trade routes and functional specialization of cities seized to be subject to indigenous decision processes or at least to be locally controlled in favour of African civilization. The alignment of the trade routes changed sidetracking rival Arab merchants and the savannah kingdoms collapsed. Most other African kingdoms fell and were unable to get up on account of this manoeuvre which benefited external interest. These events initiated the process of installing and managing capitalist economy in urban Africa.

Imperial space economy had the effect of reworking the system inside out, upside down. At its outset, negotiated control of the space economy informed the incidence of what is commonly tagged traditional urbanization. The so-called traditional urbanization period witnessed the transition in the control of space economy underpinned in practical terms with the empowerment of the trading outposts. This meant a shift in development paradigm, which saw the trading outposts metamorphose into permanent settlements and in the process took over the traditional functions of the parent enclave settlements. Having lost their symbol of civilization to a coercive alternative, a society in Diaspora emerged.

Somehow, slowly and steadily, the integrated cosmology of traditional Africa was replaced with single-minded utilitarian objectives which produced utilitarian designs for cities in Africa. The design options bulldozed cultural symbols, behaviour and beliefs that determined the system of base of traditional African cities. Cities in Africa became hybrids, an inevitable product of intervening culture and policy formulation hegemony from abroad. Beginning from the mid-nineteenth century, cities in Africa were no longer 'African cities' both in character and in function because the institutional framework underpinning them altered significantly. 'African cities' became cities in the Diaspora in their homeland.

The Portuguese in the sixteenth century founded Bissau in Guinea; Luanda, Benguela and Sao Salvador in Angola; and Laurence Marques, Sena and Mozambique in Mozambique; as well as temporarily wrestling control of Zanzibar and Mombassa from Arab traders. The Dutch founded Cape Town in 1652 and the British a number of West African ports, including Conakry, Accra, Secondi, Cape Coast and Calabar. As events that further disenfranchised Africa unfolded in early nineteenth century, which included two and a half centuries of slave trade (fifteenth–mid-nineteenth century), these towns turned out to be forbearers of colonial towns that heralded western pattern of urbanization.

Subjugation to imperial authority marked the era of colonization (Davidson 1978). Africa experienced colonization in tandem with the surge of neoliberalism in the 1930s. This marked the second phase of imperial control of space economy in the region. After the Berlin conference, the doctrine of assimilation expressed in the

French, Portuguese and Belgium territories and the paternalist philosophy of administration in the British territory introduced further controls over indigenous economies in Africa. Trade monopoly, taxing of peasants and forced labour were common in West Africa while in southern and parts of Central and Eastern Africa land was expropriated, taxes imposed and African agriculture discriminated against. The colonial system restructured peasant agriculture, introduced new administrative systems and changed the pattern of urbanization (Rakodi 1997).

The traditional socially homogeneous urban space found especially in Anglophone Africa became segregated, a transformation that was more common in Francophone Africa where deliberate efforts were made by colonial administrators to plan the urban space. Apart from redistributing the urban pattern, the combined effect of trade and colonial planning intervention altered the structure of city in divergent ways. Traditional African cities converted to hybrid cities that 'harbour hundreds of differing allegiances, smouldering in recalcitrance and intolerance, flaring up unexpectedly in hatred and mass hysteria' (Oliver 1971: p. 229). New towns, colonial towns, sprang up on territories with the greatest concentration of natural wealth and with convenient access to ports and railways (Rimsha 1976).

There were two categories of colonial towns that emerged, both of which served and are more or less still serving as suppliers of raw materials and outlets for western manufactured goods (Oliver 1971: 227). The categories found mostly in West Africa's francophone region are mostly outgrowths of existing 'new' settlements founded during the slave trade period. The other category found in the settler territories of North, East and Southern Africa by contrast were 'new towns': European in appearance, planning and organization and served by an itinerant class of landless peasants, migrant black paupers, mine boys and domestic servants (Oliver 1971: 227).

The colonial city was a pastiche of zoned functions, land uses and population. There is high indication that Africans do not feel a sense of civic identity for this category of cities. In it, according to Oliver (1971: 229), Europeans controlled the modern commercial and administrative "center (the industrial, transport and military zones) and the well-kept residential compounds for Europeans and the African bourgeoisie on the other hand controlled the old city, the (stranger's) quarters and the sprawling zones of squatters—the trespassers of desperation—scattered in the city and on its outer edges".

The incidence of independence within the 1950–1960 period did not impact the imperial space economy agenda. All it did was to engineer imperialism as administration to control the society in Diaspora. This aided the recession of planning regulation and the growth of informalization in the drive to retain the status quo in the space economy. By implication a choice is made to continue with western-oriented capitalist urbanization concept linked with market economy which upholds the principle of privatization. This was done against the option of soviet style socialist urbanization, which is built on state intervention (a critical policy objective of national governments in African countries). More than five decades of experimentation with the western concept yielded deepening poverty and low productivity.

Since the turn of twenty-first century, neoliberalism as global economic orthodoxy and meta-theory for planning has worked to sustain the resilience of imperial space economy and to reinforce extroverted economies that have limited links and relationships with the local economy. By extension neoliberalism domesticates the attributes of core periphery in the world system described in Wallerstein's dependency theory (see Wallerstein 2004). In essence what remains of Africa's control of her space economy is being torn apart, as neoliberal economic policy in global economy reinforce the instrumentality of market force through neoclassical investment mechanisms to manage economic land use development particularly in African countries.

3.2 Informality in Planning

Informality has several conceptual representations but dominantly economic in orientation. In the same vein, planning—presumably represented by spatial planning concept—has undergone dynamic changes in its theoretical meaning. The causality between informality and planning is often the focus of intellectual scholarship since the turn of the twenty-first century. Informality, which is normally perceived in the context of the informal sector, is lately extrapolated to be perceived in the planning context.

3.2.1 Notions of Informality

The rethinking of informality, which Parthasarathy (2009) presented, attempted the representation of the concept in new ways. It spotted attempts to link the concept with processes of economic liberalization (Roy and AlSayyad 2004, and Breman 1999); with global regimes of accumulation (Sassen 1990), which relates more with introverted economies; and with 'fiscal sociology' (Schrank 2009). Besides, Parthasarathy (2009) attempts to push the concept beyond into political, religious and cultural fields, to reflect on informality in terms of struggles over autonomy, self-expression, and resistance—which implicate religion and politics and their spatial expressions in unique ways.

The concept of informality reflects an opposition to various levels of organization and institutionalization (Parthasarathy 2009). Parthasarathy points out that this concept brings out several aspects of informality especially the use of public space which 'are usually characterized by a multiplicity of use simultaneously and across time, space is easily modified and made flexible to fit multiple uses, and land use change occurs according to different notions of time, of planners, of capital, of street vendors and hawkers, and of sundry other consumers of space'. According to Roy (2005), although the idea of composite space supports de-zoning principles it simultaneously reinforces the concept of informality as a land use problem at first glance. Thus it is 'often managed through attempts to restore "order" to the urban landscape or to bring it into the fold of formal markets'. Roy (2005) leaned on Krueckeberg (1995) to redirect attention to what she termed 'more fundamental issue' at stake in informality, which 'is that of wealth distribution and unequal property ownership, of what sorts of markets are at work in our cities and how they shape or limit affordability'. Therefore, informality straddles issues relat to the 'right to the city' and the 'right to property'. Watson (2009) expounded these rights with the idea of 'conflicting rationalities' which explains the logic of governing and the logic of survival. In her submission, it is obvious that the logic of survival is the central element of informality.

Informality, based on the logic of survival in extroverted economies found in the global south, hold extra implications that bother on corporate subjection, dependence and marginalization. These factors point to the survivalist informal sector as opposed to the modern informal sector, which relates to global regimes of accumulation (Sassen 1990). Therefore, conceptually, informality could be positive or negative depending on the relative position of political economies in the world system. Hence, informality is by no means a universal constant, oblivious to context (Duminy 2011: 3). The negative conceptualization tends to be dominant and reflects on the themes of usage of informality. Duminy (2011: 1) identified some common themes of usage of informality in planning literature. They include: informal planning; urban informality (Roy and AlSayyad 2004); informalization of cities (see Simone 1998); modalities of urban associational life (e.g. 'informal social networks') (Simone 2001); informal space (Roy 2009: 84); etc. These themes inform the nature and definition of informality.

At the turn of the twenty-first century there are two contrasting frames of informality—one of crisis and the other of heroism. The frame of crisis is drawn from the outcome of the concern of Hall and Pfeiffer (2000) for what they referred to as 'informal hypergrowth' cities. They argue that through migration 'some cities of the developed world are invaded by the developing world' rendering them ungovernable (Hall and Pfeiffer 2000: 129). In contrast with this language of crisis, Hernando de Soto (2000), in his superselling book *The Mystery of Capital*, presents an image of informality as 'heroic entrepreneurship.' (Roy 2005: 148). de Soto (2000: 14) held a rather institutionalist position that the 'informal economy is the people's spontaneous and creative response to the state's incapacity to satisfy the basic needs of the impoverished masses'. The frame of heroism draws from de Soto's position. Both the crisis and heroism frame consider informality within the premise of the informal sector, which highlights the element of population (people and activity).

Therefore, there are some measures of relativity in the nurturing of informality. Duminy (2011: p. 8) drew attention to the 'stratification of informalities' through the designation of certain activities as acceptable and legitimate ('whitening'), others as pernicious or criminal ('blackening'). Graphically, the informality of the powerful is whitened, that is rationalized and perhaps given legal status and that of the poor is blackened; that is condemned and perhaps punished (see Roy 2009: 87). It becomes difficult to define informality.

The definitions of informality most often point to those activities that do not closely follow the law (for example, not paying taxes) or institutionalized planning regulations (Duminy 2011: 2). These activities could be for survival or for accumulation and Duminy (2011) further explained based on Roy (2005) insight that 'informality, although often viewed as being localized in nature (a local issue), is embroiled in a vast complex network of policies, decisions and grassroots actions'. From this standpoint Roy (2005) perceived informality as a 'series of transactions that connect different economies and spaces to one another'. Within the same year and outside the informal sector domain she perceived informality as a mode of metropolitan urbanization and as an expression of sovereignty. A couple of years later she moved into the planning arena and gave the definition of informality to mean 'a state of deregulation, one where the ownership, use, and purpose of land cannot be fixed and mapped according to any prescribed set of regulations or the law' (Roy 2009: 80). This time around the definition directly attacked conventional planning following what Roy (2005) termed 'policy epistemologies' (moving from land use to distributive justice, rethinking the object of development, and replacing best practice models with realist critique), which ultimately sought new ways of viewing the urban space. Her contribution highlighted the element of space in the consideration of informality.

Roy's (2009) definition portrays the idea of informality as land use dynamics through informal processes, which determines 'urban informality'—an organizing logic, a system of norms that governs the process of urban transformation itself (Roy and AlSayyad 2004). The multiple use of space facilitates changes in uses of spaces across time, and even temporary changes are made use of in expedient and innovative ways by a range of social actors (Parthasarathy 2009: 8). This follows informal processes which reflect 'social actors seeking to make use of sanctioned, authorized or approved spatial practices to achieve a different set of objectives' (Parthasarathy 2009: 9).

It is observed that Roy's (2009) regulatory definition of informality is ironically elitist given the relativity identified in informality, which she acknowledged. Her definition is presumably targeted at protecting the poor but it unwittingly undermines the poor whose acts of informality are prone to be blackened. Informality does not constitute a weapon for the weak although it appears so on face value. Informality as a process or as end state (i.e. urban informality) is not a function of land tenure. Her definition is contestable because it is built on a questionable premise, which assumes that planning is instrumental to right to land. This assumption is not necessarily correct because land tenure is independent of planning in traditional African context, for instance in Nigeria. Land acquisition is factored on finance/affordability, irrespective of formal/informal or planned/unplanned context. In fact informal land market unlike policy-studded formal land market makes life more miserable for the poor because its operations are predicated strictly on finance. Whatever relationship informality has with planning it does not convey access to land or right of land ownership and this is stated clearly in the building permit issued to land developers. Planning, even informal planning, is concerned with development and should not be held responsible for land tenure difficulties which impact the poor. So the effort to stop planning from sanitizing the use of space does not guarantee access to land for the poor (see Watson 2009: 2270). Informality in the urban context realistically retains the poor in Diaspora regardless the state of exception that reflects the right to the city.

Informality in Africa is found to be based on the logic of survival and the rationality of survival does not observe ground rules in the use of space. This scenario represents the negative concept of informality because informality disconnects with accumulation. Therefore, the logic of survival tends to overwhelm the logic of governance as presented in Watson's (2009) argument of 'conflicting rationalities'. Watson's argument presumes that informality is beneficial to the poor. To the contrary, informality ultimately does not benefit the poor neither does it make land for instance available to the poor given the relativity or indeed the ambivalence of the concept in the context of power. Informality rather retains the poor in Diaspora and disposes of them at the approach of powerful neoliberals.

Informality may boost survival but it sustains poverty. Indeed informality is an epistemology of poverty. Thus this work argues that the rationality of survival will not enhance the productivity of urban Africa because it lacks sufficient theoretical bases for growth. Productivity in real terms requires the introversion of the space economy based on the rationality of accumulation. This mindset, which is directed at national building, should determine urban planning in Africa. Hitherto the disposition of planning ruled with informality is built on the neoliberal principles of private profitability. This is why the entire idea of informality in planning is more or less a political necessity submerged in the politics of power. Thus for all intents and purpose informality is not necessarily a planning requirement.

3.2.2 Notions of Planning

Ideally there are two notions of planning: statutory and spatial planning. But since 1960s a quiet revolution is sweeping through planning concepts. Most of the inputs for the reconsideration of the planning concepts are Euro-based coming from European Union (EU), European countries and the United Kingdom. Inputs from UN-Habitat are ubiquitous but largely informed by the intellectual discourse in European context especially in EU countries. The very little heard from developing countries are also Euro-centric. The planning concept altered as it passed through three distinct planning periods, namely, premodern, modern and postmodern planning era. The summary position, given Euro-centric perception, is that statutory planning associated with traditional land use planning is considered outdated while spatial planning is still undergoing momentous regeneration.

As a theoretical concept spatial planning diverts attention from the statutory attributes of planning to the broad spectrum of non-statutory attributes that apply on a wider scale. As an activity it takes different forms in different contexts, depending on institutional and legal context, or variations in planning cultures and traditions

(Healey 1997). So in terms of coverage of issues and scale of operation, spatial planning is a more integrated concept. Trends in early twentieth century witnessed the increase of its application in economic planning, resource management, conservation and its initial experimentations in modern urban planning.

Conceptual models of urban spatial planning started with strategic spatial planning, which focused on the management of land use change (Todes et al. 2010: 416). It is assumed that strategic spatial planning addresses many of the problems of old-style or statutory master planning although much depends on the actual ethics and values which the plan promotes (whether or not it promotes and enforces sustainable, inclusive cities), the extent to which the long-term vision is shared by all (and not simply dominant groups or individuals), the extent to which a stable and enduring consensus on the plan can be achieved, and the assumptions about the role and nature of space and spatial planning within the plans (Healey 2004). Gradually, it transited to comprehensive spatial planning with enlarged content, which looks at the spatial dimension of all strategic policies and aims at integrating and coordinating all space-consuming activities in a geographic territory. In other words, it goes beyond traditional land use planning to bring together and integrate policies for the development and use of land with other policies and programmes which influence the nature of places and how they function. The inherent procedural/methodological renaissance and scoping define the point of departure for comprehensive spatial planning models (Albers 1986: 22). This meant increased understanding of local contexts and formulating guiding frameworks that inform land development and management (Landman 2004: 163). Thus participatory process is encouraged in spatial planning. But as informality gained momentum in the participatory process, spatial planning lost considerable impetus as tool for practical application in direct intervention in geographic space and increasingly found expression in economic planning, resource management and conservation.

In neoliberal context, the participatory process is the entry point of informality in planning. The synthesis of informality and planning in which informality is an internal element in planning leading to participatory planning is the current vogue. Meanwhile the attributes of participatory planning point to non-statutory, informal, locally oriented set of activities that derive from needs approach in community development planning. It highlights the process-oriented approach to participation where stakeholders are involved in decision taking. Therefore, participatory planning is defined as an interactive set of processes through which diverse groups and interests engage together in reaching for a consensus on a plan and its implementation (Gallacher, n.d.). It operates with the principles of mediation and as always the participatory process presents a trade-off between efficiency and inclusiveness depending however on the level of participation anticipated. Time pressure, the needs of the community, the skills and experience of those participating, and the nature of the intervention, among other factors, all help to dictate the actual shape of the process (Rabinowitz 2013). In Africa the deliberations in the process are presumably compelled to remain within the survivalist level due to obvious limitations factored in top-down mobilization of participants, informalization and manipulative tendencies.

The idea of strategic spatial planning is currently staging a comeback after South Africa experimented in vain with roughly seven alternative process-oriented approaches to spatial planning within the 1980s–2010s periods, each with an average lifespan of 5–6 years (Okeke 2015). Nevertheless South Africa's search for appropriate spatial planning model reached some milestones in three directions: first, linking spatial planning with infrastructure development or sectoral planning, second, confirming that recourse to neoliberal ideology had the effect of politicizing planning beyond the traditional levels of politics in planning, and third, realizing the strategic need for statutory planning, strong leadership, and detailed plans with which to direct development as prerequisites for effective planning intervention (Okeke 2015). These are hard facts ordinarily in the rhetoric of planning in neoliberal context will not be easy to accept.

However, there are outstanding issues yet to be resolved particularly finding a synthesis of planning paradigms that will simultaneously deliver spatially equitable economic growth and shape the cities. To this end strategic spatial planning re-emerged since 2000 to 'promote more compact and integrated cities, and to redress patterns of inequality of the past' in Johannesburg (Todes 2012: 158). In its new disposition, which is somehow not very visible in literature, strategic spatial planning is thought to be 'more flexible (focusing on directional guidelines and key strategic interventions); go beyond land use plans to bring together sectors and institutions; involve forms of collaborative planning, including a range of stakeholders and wider levels of participation (UN-Habitat 2009); and link to implementation through projects and budgets' (Todes 2012: 159). Moreover, it portends 'to link spatial planning and infrastructure development through a growth management strategy' (Todes 2012: 158). The elements in the package of strategic spatial planning initiatives in South Africa are: the 2006 Growth and Development Strategy (GDS), statutory IDP and statutory Spatial Development Framework (SDFs) first published in 2003 from which more specific Urban Development Frameworks (UDF) have been developed (Todes 2012: 158).

3.2.3 *Informality and Planning: Notions of Relationships*

The consideration of informality from economic and planning point of view varies. In the three positions Owusu (2007) crystallized (mentioned earlier) informality is linked with the provision of employment hence it focuses on people and activities —specifically the labour force. On this account, it attracts commendation even from scholars with neo-Marxist inclination such as Manuel Castells and Alelandro Portes (see Castells and Portes 1989), Martinez-Vela (see Martinez-Vela 2001), (see Portes and Walton 1981), etc. Viewed from planning perspective, informality relates to the distribution of people (population) and activity in space. The element of space distinguishes the planning perspective hence its focus on space economy. In the African context, informality relates directly with extroverted space economy, which supports exploitative relationship with capitalist production and distribution. The point of departure here is the use of space for survival. The space element is brought to bear in planning perspective for considering employment.

As the nature of informality evolved, two contending models of relationships between informality and planning emerged: first, informality as external element to planning that is informality as an object of planning, and second, informality as internal to planning that is informality as a function of planning. The former category advocates informality for planning. This model accepts that planning is for the sake of informality. Informality is the reason why we plan. This model is associated with Ananya Roy and Oren Yiftachel. There are two aspects of the second category of relationships: first, informality with planning that is accepting informality as perfect model for planning. This model is built on participatory planning principles and demands the denial of the precepts of formal planning. The model is associated with Venessa Watson, James Duminy, Nancy Odendaal, and to some extent Ananya Roy, etc. Second, there is informality in planning, which indicates strong unity between informality and planning. It accepts that informality is the only instrument of planning. In other words informality is the central element of planning—planning without a plan. This model is closely associated with neoliberal planning school, which does not have overt followership except perhaps UN-Habitat. Both models are similar but not identical; they are both built on participatory principles but the latter is more predisposed to market force metaphor in planning. Meanwhile the three models of relationship identified are built on the institutionalist perspective of informality in the management of space economy.

One of the themes of usage of informality is the concept of 'informal planning' (Duminy 2011). Informal planning constitutes spatial planning and implementation tools that have no statutory binding force. Their integration into formal urban land use planning for example in Germany is laid down by the Building Code. The Building Code provides that development concepts and other urban development plans adopted by the local authority are to be taken into account in preparing land use plans. This system of planning is limited to local council and it is used to generate informal master plans that are binding only within the jurisdiction of the council. No formal procedures are required to prepare these plans but it is mostly oriented towards local conditions as a continuous process with stages that are not strictly chronological. Participation is voluntary although involvement of public and private authorities has become the normal practice. Informal plans elaborated for the entire community deals with social, cultural and economic demands on the settlement area. It helps to prepare political and administrative decisions and permit conflict resolution presumptuously ahead of formal planning.

Meanwhile there are two trends that crystallized in the analysis of the notion of relationships between informality and planning. The first is the synthesis of informal and formal planning to beget the art of urban design. This explains why planning to a very great extent is related to urban design in developed countries. In other words, planning in developed countries, often conceptualized as strategic spatial planning, has not completely relinquished its statutory land use planning qualities. The second trend attempts to portray informality as a function of planning. The exponents of this notion seek to reinvent planning with the instrumentality of participatory process but sufficient theoretical foundation is yet to be built to establish their informal planning model.

Informal planning tools such as Sectoral Development Planning, Sub-area Development Planning and so on are held to be much more flexible than formal urban land use planning. They apply at neighbourhood level for existing urban areas suffering from deficiencies, and primarily used for urban extensions (Pahl-Weber et al. 2009). Due to limited information the performance of these planning tools is not clear. But admittedly it is quite an ingenious and innovative strategy because it provides ample opportunity for informal inputs in formal planning.

However, informality in planning is a classic ideology that lacks relevance in addressing substantive spatial integration for Africa. There is need for a strategic withdrawal of the binary opposition that is deployed in the interpretation of this phenomenal trend. The analysis of rationalities should seek to integrate spatial equilibrium and spatial determinism in the context of territorial planning. In other words, the complementarities of economic and spatio-physical bases of growth in a formal context should be emphasized in the management of the space economy.

3.3 Informalization of Cities in Africa

3.3.1 *Contextualizing Major Arguments and Presumptions*

The informalization of African cities is remotely linked with neoliberalism vis-à-vis neoliberal planning theory, which supports informality in planning. Neoliberalism as a development ideology prospects to retain Africa in dependent capitalism therefore it supports the strategy of sustaining African urban economy with the survivalist informal economic sector. The institutionalist school, which is positively disposed to neoliberalism, sees the informal sector as instinctive and creative response to excessive and inappropriate regulation by the state (de Soto 1989; World Bank 1989). This viewpoint has become very influential in policy circles and has been incorporated into the work of neoliberal economists, policy advisors and non-governmental organizations, partly because it conforms to the global push for neoliberal and supply-side economics (Rakowski 1994; World Bank 1989: 10). This provides the background for the receptive attitude towards informality. It is on this premise that the informalization of African cities is rationalized in favour of the exploitative mode of integrating Africa into the global economic system. Thus the neoliberal argument for the informalization of African cities leans towards peripheral capitalism as stated by Amin (1974), and Beinfeld (1975) and also to Cardoso and Enzo (1969) perception of dependent capitalism in the world system.

A symmetrically opposed argument is found in neo-Marxist perception, which is sceptical about the efficacy of the informal sector to benefit the poor. The neo-Marxist perception maintains that the poverty of the informal sector results from its exploitative relationship with capitalist production and distribution. This aptly explains the survivalist informal sector in Africa and the picture gets clearer against the backdrop of African renaissance. The contextualization of the sector from a substantive viewpoint supports the alternative argument that the informalization of African cities, which derives from informal sector argument presented by institutionalists, is antithetical to African renaissance. Also coupled with the form and function principles of sourcing spatial integration, currently undermined in neoliberal planning context, it is further argued that the informalization of Africa cities lack theoretical bases for growth.

In considering the alternative contextualization of the informalization of African cities, the mission to pull Africa out of dependent capitalism is presumed. What this means is that the structures that maintain the status quo needs to be dismantled. Thus the mindset developed under the influence of institutionalist point of view, which supports the receptive attitude towards the informal sector and informality in planning, need to be redressed with some elements of neo-Marxist philosophy. The renewed mindset that is anticipated will provide the platform for downsizing the impact of informality in the development of the space economy in Africa.

The informalization of African cities prevails under notable presumptions. Regrettably, African planners and their planning systems tend to pay primary attention to cross-cutting issues in planning such as climate change, global warming, environmental degradation, inclusiveness, gender issues in planning, etc., that focus on attaining urban quality rather than addressing core issues of urbanity, which deal with determining appropriate urban form for the delivery of growth visions. In other words, remedying the harm the delivery of neoliberal economic orthodoxy bear on African cities tends to absorb the attention of African planners and their planning perspectives. The alternative function of using planning instruments to deliver development ideology seems to elude the intellectual base of the African planner. This explains the presumption that arrogates attention to financial mechanism as a major determinant factor for the delivery of development ideologies. But to the contrary, the trend of events indicates that the delivery of imperialism, capitalism and the prevailing neoliberalism are sought primarily with planning instruments. Financial mechanisms play a facilitating role. Thus, the abuse of formal planning instrument was used to deliver capitalism and the current neoliberal planning theory, which crusades informality in planning, is being used to deliver neoliberalism. The informalization of African cities in the context of neoliberal planning confirms the anchorage of neoliberalism in African space economy.

Neoliberal planning nurtures the informalization of African cities. This event is premised on pseudo-urbanization where tertiary sector rather than productive economic activities stimulate growth. The tertiary sector assumed survivalist informal sector orientation following the structural adjustment programme of neoliberal economic policies in the 1980s. The informalization of human systems gained momentum courtesy of pro-poor and gender arguments, especially as it relates to land tenure and access to land. The informalization of the use of space driven mainly by informal housing found expression in urban sprawl and informal settlement development, mainly at the urban fringe. The urban sprawl derives from the processes of succession in urban ecology: unplanned, uncontrolled, and uncoordinated single use development that does not provide for a functional mix of uses and/or is not functionally related to surrounding land uses and which variously appears as low-density, ribbon or strip, scattered, leapfrog, or isolated development. This is responsible for social–physical phenomenon of land use segregation and the related ecological processes of succession and dominance that leads to an increasingly homogeneous and specialized community, the effects of which have far-reaching consequences in terms of survival probabilities (Edwardo 1990: 143). These elements, resulting in the proliferation of peripheral slums of despair, characterize the informalization of African cities.

At the moment lots of persuading arguments from sociological perspective, which culminated in advancing informal planning, tend to provide justification for legalizing informal development. This trend that holds true for transient and capitalist systems is pronounced in Albania and Greece for instance. Similar arguments have diffused into Africa. Without denying the so-called benefits of informal development, which is not consistent with modern economic systems, this discourse differ in principle with this trend in Africa. Informality is a response to inefficiency of a state responsibility in relation to good land administration and urban management, or to housing policies, to economic development and provision of job opportunities. Issues of sprawl, suburbanizationeutrophication, etc., which attend to informal development, present formidable barriers for sustainability and holds implications for climate change due to stress factors. Incidentally, legalizing informal development neither removes these barriers nor creates opportunities to remove them.

3.3.2 Diagnosis of Growth Potentials

There exists a very strong impression in African studies that the economy of African countries is located at the periphery of international capitalism. As mentioned earlier these economies are structured to meet the needs of the capitalist system (Roberts 1978; Santos 1979; Lowder 1986). Therefore, through various mechanisms such as the control of commodity markets, the activities of the multinational companies and their subsidiaries, the control of the source of credit and interest rates, and syphoning of resources from the indigenous society, African countries are made to assume a dependency status (Chukuezi 2010). Studies conducted in the 1970s by economic anthropologists such as Meillassoux (1972), Wolpe (1975), and others (e.g. Clammer 1975) indicated that subsistence enclaves were being deliberately preserved as labour reservoirs in Third World countries for the modern sector (Portes 1983: 171). This framework provides an insight to the insurgence of survivalist informal sector operators in African cities in the 1980s most probably as prelude for the delivery of Euro-American mercantilism. By the 1990s the informalization of the human system has become prevalent in Africa.

The informalization of African cities or as Bibangambah (1992) put it, the progressive 'ruralization' of urban areas, got underway with the wave of recourse to informal approaches to manage urban systems in space. The informal economic sector which drew from these approaches was established to be dominated by subsistent retail trade activities on foreign goods. These activities are conducted mainly by migrants from rural areas. Steady streams of this low-skilled labour into the cities cause the urban informal sector to expand due to the limited employment capacity of the formal sector (Ishengoma and Kappel 2005). The growth of informal sector operators relates directly to rapid urbanization as neoliberal and trade liberalization policies provide impetus. But the per capita productivity of large proportions of migrant population who are absorbed in the survivalist informal sector, tend to dwindle due to underemployment (Onyenechere 2011: 63).

At the outset, around 1980s, the attitude towards the informal sector was quite repressive in Africa. Progressively this attitude transited to become much more pragmatic and promotional. Chukuezi's (2010) reflects this attitudinal change thus:

> more than ever, policy-makers in developing countries are recognizing the vital role of the informal sector as a mainspring of vitality and diversity in the urban economy; as a leading provider of jobs for first-time job seekers, low-skill workers and migrants from the rural areas; as a proving ground for entrepreneurial talent and as a source of skill development.

Governments rather than harass or demolish informal sector areas of operation now encourage and give them some sense of direction. Apparently the main policy challenge is how to support and regulate the urban informal sector in order to promote employment, productivity, and income for the poor, and at the same time ensure a safe, healthy and socially acceptable environment (Nwaka 2005). The policy dilemma, which is yet not widely acknowledged, appears to be how to contain the adverse environmental impacts of many of the activities of the urban informal sector without disrupting livelihoods, and causing social distress (Nwaka 2005).

The receptive attitude and policy change presumably is either factored on growth potentials of the informal sector or perhaps on the potentials of the sector to manage poverty. The status of the sector in the local economy escalated within the 1970s–1990s period. Then in 1993 the sector typically added between 20 and 30 % to African GDP and real GDP growth tended to be positive apparently in direct contrast with negative growth in other economic indicators, particularly inequality increases, per capita GDP declines, and poverty increases. The scenario since the turn of the century has not significantly changed is the simple summary of the monumental institutional database on economic indicators in global economic analyses.

There are about three points of departures in the diagnosis of the growth potentials of the informalization of African cities. In the first instance, the declining urban productivity associated with the African region is inevitable given its survivalist informal economic sector, which caters to its own needs and demands (food, housing, transport, etc.) (Hicks 1998). Often the informal enterprises in the system are left to interact with precarious formal enterprises run by the public sector. The sector, unlike modern informal sector, conceptually has no provision for accumulation and under this condition the sector cannot provide the African region reasonable theoretical bases for growth. The benefits of the positive GDP growth rate in the region apparently accrues to the manufacturing formal sector in developed countries through the mechanism of trade liberalization. Without formal productive enterprises with prospect to support forward linkages with the informal sector the productive economic system in Africa cannot generate growth and accumulation. The commercial landscape in Africa seldom contains requisite entrepreneurial enterprises and where they are available they are endangered due to unintended effects of policy initiatives that encourage the survival strategy of informal economy.

Second, the operations of the informal sector are contained within the physical formal region, i.e. the city. In Africa the polarization of skilled and low-skilled labour to cities is not dependent on the availability of industries in the cities. It is more or less a response to geographical determinism generated by distorted incentives to exploit subsidies (the World Bank asserts) rather than the attraction of functional cohesion of interdependent productive units indicated in economic formal regions. This scenario has implications for connectivity within the functional region, i.e. the urban economic region. Links and relationships that determine functional flow best provided by large scale formal firms are lacking given the preponderance of less integrative informal Small and Medium Enterprises (SME's) especially the retail trade category that dominate the commercial landscape of Africa. The legacy of sprawl and suburbanization, which terminates in slum formation, exacerbates the problem with functional flow of business activities. This draws attention to the direction of causality between slum formation and informality in planning ruled by market force metaphor. The informalization of African cities is submerged in this causality and therefore cannot anticipate spatio-physical integration and more so given the total incapacitation of the form and function principles of planning rationality. So, functional flow lacks expression anyhow in the survivalist manoeuvres that sustain the economy.

And thirdly, this work argues that the informalization process disposes African cities to the delivery of the expansionist programme of external economies. The process usurps the authority of local institutions to control the economy. A symmetrical dichotomy develops in economic and spatio-physical terms commonly designated in planning literature as dualism of the economy. The components of formal and informal economies find expression in segregated space—the urban and rural areas. As the informal economy grows the formal economy shrinks thus creating a scenario where the dwindling integrative element of the mostly urban-based formal economy capacitates discrete informal sector operations to generate disconnect between the urban and rural economy. The urban-based modern sector disconnects with the local economy and disables the Christaller (1933) core-periphery theory of growth. Such extroverted space economy is a common feature in most African countries. This leaves the local natural resources redundant in favour of marketing in foreign goods as well as resource marketing, courtesy of Euro-American mercantilism. Indeed, it disconnects with the exportable product of the city which deals with the internal generation of goods and services for growth processes that implicate development. Growth is not likely to issue from this structure because its mechanism does not comply for instance with the basic principles of export-base theory of city (urban) growth in economic terms put forward by Tiebout and North in 1956. This theory amongst others treat the growth of cities as a process implicated in functional flow in space and not as isolated events found in contemporary African cities that is undergoing informalization.

Since 1990s when informalization gained momentum, the development of growth theories leading up to the new economic geography of neoclassical theories is consistently silent on the factor of informality in city development. Neoclassical theories developed within the realm of economic growth fundamentals that sought economic bases of integration. But it differed in orientation compared with the preceding structural and institutionalist theories. The neoclassical category of theories was introverted, thus signalling endogenous growth theories, which argue that the growth of a region is internal. Although the theory holds this critical position against the informalization strategy, it commits to paradigm shift in planning in the attempt to address efficiency intervention in the name of equity and in the light of market failures. Compatibility issues arise here and manifest in the rejection of planning rationality as instrument for spatial integration. However, planning had to contend with the synthesis of efficiency and equity which is the legitimate concern of endogenous growth theories in early 1990s. It is not clear if solution to compatibility is linked with the theory of new economic geography in late 1990s. This theory reconsidered geographic space and introduced the spatial dimensions of regional growth and trade. Shift in emphasis to spatial integration is once more noticed thus indicting the informalization process.

3.4 Basic Scenario of Planning in Select African Countries

The submissions in this subsection starts with a question: what if the planning scene in African countries is tending towards neoliberal participatory planning (under the influence of imperialism as international administration) for purposes of sustaining the imperial space economy, which retains the dependency status of Africa in a politicized neoliberal globalization? Otherwise, how do we explain the role, or why are the so-called development partners instrumental to the transition from formal planning for spatial integration to a wide range of neoliberal participatory planning perspectives (reminiscent of investment planning), which are focused on project development for private profitability? For purposes of assessing the planning scene in African countries, planning instruments and the turn of events in fourteen thematic areas that are related to the subject and object of planning are examined in ten select African countries. An educated summary of impressions are gleaned at country profile level to establish the basic scenario of planning in select countries. The iterative database generated is presented in Table 3.1.

The summary of impressions of the planning scene in Africa derived from the iterative database from select African countries is presented in Table 3.2.

3.4 Basic Scenario of Planning in Select African Countries 51

Table 3.1 Iterative database of selected African countries

Thematic areas	Sources: Line of argument (country level)		
	DRC	Angola	Mali
Country profile (growth indicators)	• Gained independence five decades ago, spatially double the size of most African countries and graded second in landmass; operates three-tier territorial divisions with dominantly poor, large and growing population graded third position that engage mainly and increasingly in agricultural activities • The weak national economy manifests increasing GDP, unstable GDP growth rate, declining per capita productivity, growing inequality, increasing level of urbanization with the incidence of megacities although with relatively medium urban growth rate • Apparently active in regional networking and anticipates to consolidate national unity presently undermined by successive wars *References* ADB/ADF (2009: 17), McGranaham et al. (2009: 43), UN-Habitat (2010), Gumilla Olund Wingqvist (2008: 2), Putzel et al. (2008: 7), UN-Habitat (2010: 181)	• Gained independence approximately four decades ago, spatially half the size of the DRC and graded fifth in landmass, operates two-tier territorial divisions with about half of the relatively small population that is graded fourteenth position within poverty level with more than half the labour force engaged in informal agricultural activities • The extroverted national economy grew phenomenally in mid-2000 mainly on account of petroleum related industrial activities with corresponding increase in per capita productivity although inequality grew as well perhaps indicating lopsidedness in economic growth, itself slowing down at reduced growth rate • Higher urbanization level is anticipate although megacity syndrome is a remote possibility at the moment, however the primacy of Luanda is evident • Apparently active in regional networking and anticipates a new era of peace, reconstruction and reconciliation *References* McGranaham et al. (2009: 43), UN-Habitat (2010: 174, 181), Sousa (2002: 97)	• Gained independence five decades ago, spatially almost same size with Angola and operates multiple levels of territorial divisions for a relatively small but increasing population half of which are poor and engage mainly in agricultural activities • The impoverished national economy which is agriculture driven is improving gradually at a moderate growth rate and positive per capita productivity however with clear indications of asset poverty inequalities • Urbanization level is low although its growth rate is high but there is no indication of megacity development • Actively involved in regional networking and anticipates to control the destiny of its civilization *References* Farvacque-Vitkovic et al. (2007: 25), Djiré (2006:1), McGranaham et al. (2009: 43), Yousif (2005: 58), Farvacque-Vitkovic et al. (2007: 1, 35–38), Yousif (2005: 58), GSDI (2010: 25), UN-Habitat (2010: 53)

Table 3.1 Iterative database of selected African countries

Thematic areas	Sources: Line of argument (country level)		
	DRC	Angola	Mali
Urban planning machinery	• More concerned with environmental policies with backup legislation specifically focused on forest management as such statutory zoning regulations guide land use planning • Available agencies are committed to land affairs, agriculture development and environmental management; managing the large landmass and the forest most probably informed the environmental orientation of the planning machinery *References* FAO (2009a), BDA (2007: 1–59)	• Operates decentralization policy that is more concerned with redistributing the national economy although they have a crop of planning laws enacted mid-2000 that deals directly with urbanization and urban development perhaps in anticipation of the spatial implications of their decentralization policy and some of the legislation specify regulations on physical planning • There exists a well-developed range of agencies concerned with urban planning and urbanism, enough to effectively deliver in space the decentralization policy *References* Documents: • Land Use Planning and Urban Development Act (no. 3/2004 of June 25) • Angola's 2004 Land Law • 2004 Law of Territorial Planning and Urbanization (Lei do Orderamento do Territorio e do Urbanismo, Lei 03/04, 25 June 2004) • 2007 Land Law Regulations; Environment Framework Law in 1998 (Lei de Bases do Ambiente), No. 5/98 of 19 June • Basic National Planning System Law No. 20/11 of April 14, 2011	• Decentralization policy backed by legislation, squarely concerned with territorial divisions; land use planning is a directorate function most probably to support ministerial urban development functions; planning activities is guided by regulatory urban planning documents in Francophone countries *References* Farvacque-Vitkovic et al. (2007: vii, 28/52), World bank (2002c), Attahi et al. (2009: 35)

(continued)

3.4 Basic Scenario of Planning in Select African Countries

Table 3.1 (continued)

Thematic areas	Sources: Line of argument (country level)		
	DRC	Angola	Mali
Urban form	• Introverted urbanization leading to city primacy with Kinshasa as centre of attraction; the urban form manifest sprawl spatial pattern of expansion aided by dispersed land use distribution especially at the periphery areas due to processes of succession, although some measure of gentrification is also manifest and is responsible for smart growth occasioned by the activities of real estate agents *References* UN-Habitat (2010: 174, 175, 178, 182, 189)	• Introverted urbanization characterized by city primacy and informal settlement development at the fringe areas hence both processes of gentrification and succession are manifest; the urban form is cryptic with very complex urban patterns and dispersed land use distribution associated with informal smart growth operations resulting in uncontrolled densification *References* Jenkins et al. (2002: 118,124), UN-Habitat (2010: 174,175), Cain (2004), Development Workshop (2005: 162)	• Sever introversion of urbanization responsible for acute primacy of Bamako the capital city; on the one hand the urban form manifests extensive sprawl that features dispersed distribution of heterogeneous land use patterns lead by qualitative urban growth, on the other hand downtown areas experience densification due to city compaction which ensued in late 1980s *References* Farvacque-Vitkovic et al. (2007: 2–4), Kilroy (2008: 13–15,32)
Transboundary problems	• Sprawling informal expansion of the urban system into suburban areas in reaction to the suburbanization of poverty that derives from extroverted dualistic urban economy *References* UN-Habitat (2010: 177, 189, 200), Putzel et al. (2008: 3)	• The enclave-style economy coupled with insecurity in the hinterland areas informs the proliferation of unproductive informal settlement development in the peri-urban areas, leading to uncontrolled spatial spread of the primate city of Luanda to absorb nearby towns *References* UN-Habitat (2010: 177,200), Jenkins et al. (2002: 121), Cain (2004), ADB/ADF (2010: 6)	• Informal physical expansion of Bamako into thinly populated outlying areas as extroverted urban economy drives the impoverished society into survivalist informal sector that disconnects from the husbandry of local resources *References* Farvacque-Vitkovic et al. (2007: 4,37), Kilroy (2008: 13,32), World Bank 2002c), UN-Habitat (2010: 110–116), GSDI (2010: 25)

(continued)

Table 3.1 (continued)

Thematic areas	Sources: Line of argument (country level)		
	DRC	Angola	Mali
Levels of spatial/economic development planning	• Apart from the master planning experience in the 1950s and pilot zoning at regional level there were no formal spatial planning effort in DRC until 2005 *References* BDA (2007: 1–59), Doc(s): National Development Plan (2007–2012)	• Stratified planning frameworks across territorial levels: action agenda at local level, master planning at regional level, and development framework at national level *References* Doc(s): Sustainable Development (2009–2013), National master plan for Angola's coastal zone; 2006 Downtown Action Agenda: Angola, Indiana Downtown Revitalization Plan 20/20	• Applies land use planning across board *References* Farvacque-Vitkovic et al. (2007: 320,44), World bank (2002c)
Urban planning instruments	• Planning initiatives cut across the various levels of territorial division and are invariably committed to the design principles of zoning and land use regulation expressed in master plans and or structure plans	• Statutory expertise oriented master planning with design considerations for territorial planning initiatives	• Strategic planning through participatory process most probably a recent trend induced by development partners *References* World Bank (2002c), BNETD (2001)
Planning education facilities	• Apparently not committed at the moment to planning education given the dearth of planning education facilities	• Dearth of a planning education facility is identified as it is the case in DRC. Perhaps the war situation in these countries could be a contributing factor	• Affiliated with the facility in Lomé shared by Francophone countries in addition to a tertiary facility in Bamako. Both facilities are not considering serious action towards planning education *References* Attahi et al. (2009)
			(continued)

3.4 Basic Scenario of Planning in Select African Countries

Table 3.1 (continued)

Thematic areas	Sources: Line of argument (country level)		
	DRC	Angola	Mali
Professionalism (urban planning)	• No formal structures in place for managing professional planning practice	• No formal structures in place for managing professional planning practice	• No formal structures in place for professional planning practice
National Urban Development Strategy (NUDS)	• Development corridors strategy reflect concern for space economy most probably left to develop informally *References* ADB/ADF (2009: 1, 13)	• No definite strategy besides development objective *References* Farvacque-Vitkovic et al. (2007: vii)	• Growth centres and development axes strategy most probably lead by market force *References* Farvacque-Vitkovic et al. (2007: vii)
Development cum transport corridors/transnational corridors	• Wide ranging network of development corridors with prospects of developing transnational mega urban region (Kinshasa–Brazzaville) *References* UN-Habitat (2010: 12), de Beer (2001: 6), Thomas (2009)	• Angola's national corridors are more or less logistic corridors most likely prone to serve as conduits for the spread of market-space economy of external economies *References* ADB/ADF (2010: 9), UN-Habitat (2010: 13), de Beer (2001: 6), Thomas (2009)	• Littered with transnational transport corridors that are more or less transit channel in dependent economies commonly associated with African states. Consideration should be given more to the growth triangle between Mopti–Bamako–Sikasso *References* Farvacque-Vitkovic et al. (2007: 3), Thomas (2009), ADF (2010: 2); Briceño-Garmendia et al. (2011: 5)
The informal sector	• Significant in the husbandry of national economy but has the externalities of generating informal settlements that swells slum population at the urban fringe areas. Reformist perception of the sector informs the receptive attitude towards it hence the liberalization reforms in the 1980s	• Although it contributes significantly to the national economy it was until mid-2000 regarded as illegal activity and exclusionary attitude mitted towards it. Reformist perception kept it afloat to service mainly retail trading entrepreneurship	• The resilience of repressive attitude towards the sector in spite of its significant impact on national economy is remarkable perhaps it is not unrelated to survivalist type of the sector that is manifest in the country and its capacity to degrade environmental and urban quality

(continued)

Table 3.1 (continued)

Thematic areas	Sources: Line of argument (country level)		
	DRC	Angola	Mali
	References Putzel et al. (2008: 3, 5–7), Thomas (2009), Xaba et al. (2002), UN-Habitat (2010: 80)	*References* Xaba et al. (2002: 6), Jenkins et al. (2002: 121), Wekwete (1995), Cain (2004)	
Paradigm shift in spatial and statutory planning	• Recovered in 2000 from more than two decades of urban planning vacuum and maintained fate with land use master planning as the best practice instrument to manage their vast forest resources	• Familiar with master planning but since 1990s attempts to imbibe at least in principle participatory process in a new master planning approach particularly at the local level	• Trend and exogenous influence in the 1990s made possible the inception of the principles of participatory planning. It is not clear how successful the UMP participatory approach used in 2000 for the City Development Strategy in Bamako
	References BDA (2007: 1–59)	*References* Development Workshop (2005: 161), Cain (2010: 152)	*References* Attahi et al. (2009: 18, 34, 36), World Bank (2002c), BNETD (2001)
NEPAD	• Remains committed especially with the agriculture network programme and shares from AU/NEPAD action plan	• Not very active in NEPAD activities	• Fairly active and recently got involved with the agriculture network programme of NEPAD
	References Yousif (2005)	*References* NEPAD (2002: 235), Yousif (2005)	*References* AMCEN (2008: 10), UN (2004: 23), Yousif (2005)
Urban planning consultancy services	• Foreign consultants dominate planning consultancy services	• Not much of consultancy activities but the little that was done was guide by foreign input	• Foreign input is evident but local effort is active
	References Miscellaneous (Internet) material	*References* Miscellaneous (Internet) materials	*References* Attahi et al. (2009: 34), World Bank (2002c)

(continued)

3.4 Basic Scenario of Planning in Select African Countries

Table 3.1 (continued)

Thematic areas	Sources: line of argument (country level)		
	Egypt	Senegal	Kenya
Thematic areas	Sources: line of argument (country level)		
	Egypt	Senegal	Kenya
Country profile (growth indicators)	• Gained independence precisely nine decades ago; spatially larger than Nigeria that is graded tenth in landmass; operates one-tier territorial division with a large increasing population that engage in service delivery and three-quarters of them are above poverty line • The national economy which is service driven is above average and stable with good and equitable per capita productivity • Urbanization level is midstream against the backdrop of remarkably low urban growth rate • Actively involved in regional networks and prospects to maintain sustainable growth inclusively *References* McGranahan et al. (2009: 43), Madbouly (2009: 7, 18), Kessides (2005: 71), EEAA (2007), UN-Habitat (2008: 45), UN-Habitat (2010: 26, 53, 70, 73), OSISA and Oxfam (2009: 64)	• A spatially small state that gained independence about five decades ago, has an equally small population that engage mainly in agriculture and somehow maintains low level of poverty • Incidentally the economy is service driven contrary to the employment structure and as per capita productivity increases inequality increases also, perhaps larger percentage of the population are not responsible for the GDP and this most probably explains the unstable and yet low GDP growth rate • The level of urbanization is average and prospect of growth is low • Their involvement in regional networking is remarkable with prospects of developing a viable economic entity *References* World Bank (2002), McGranahan et al. (2009: 43), Kessides (2005: 71), ADB/ADF (2010: 3), UN-Habitat (2010: 53)	• One of the spatially medium-sized states segregated into three territorial divisions that gained independence five decades ago with relatively large growing population compared to Senegal, three-quarters of which engage in agriculture as in Senegal and a little less than half of the population (down from a higher value) is below poverty line as at 2006 • Again contrary to the employment structure the national economy is drive by the service sector in the context of increasing per capita productivity that expectedly results in increasing inequality • Urbanization levels fluctuates though much is not expected in spite of the high but dwindling urban growth rate • Maintains remarkable involvement in regional networking and prospects to develop economic zones in anticipation of a globally competitive and prosperous country *References* Johnstone (2004: 45), Republic of Kenya (2002: 11), McGranahan et al. (2009: 43), Yousif, (2005: 58); Yousif (2005: 58), UN-Habitat (2010: 53), Daniels (2010: 19), ETC (2006: iii), UN-Habitat (2005/6: 4), OSISA and Oxfam (2009: 64)

(continued)

Table 3.1 (continued)

Thematic areas	Sources: line of argument (country level)		
	Egypt	Senegal	Kenya
Urban planning machinery	• The new towns and informal settlement upgrading policies are operational here under a set of planning legislation enacted in the 1970–1980 period which are in favour of master planning but lately in 2008 a new law instituted participatory planning; institutional provision for planning is evident but the documentation of planning standards is not evident as the planning perspective transits *References* del Rio Garcia Luis et al. (2010: 22), Ibrahim (1993: 149), Madbouly (2009: 37)	• Policy provisions are available for upgrading and legalization of squatter settlements and two presidential provisions backup this policy; land legalization is the main focus of attention considering the painfully limited landmass unlike most states with limited landmass such as Singapore, Israel, Kuala Lumpur that characteristically focus on land use planning policies; the composition of agencies indicates that planning is a residual matter and it is not clear why it is so because the prospects of economic growth anticipated for the country has implications for form *References* World Bank (2002), Attahi et al. (2009: 35)	• With a crop of substantive land use planning related policies and legislation the planning machinery is stable and more so with planning agencies at various levels of territorial divisions fortified with active and organized civil society; government adoptive bye-laws of late 1960s subsists for planning regulations *References* UN-Habitat (2010c: 17), Oyugi and K'Akumu (2007: 101), UN-Habitat (2005: 4), Mabogunje (1990:160), Berrisford and Kihato (2006: 24), Omwenga (2001: 1), UN-Habitat (2005: 56, 58)
Urban form	• The traditionally introverted urbanization sustains the primacy of Greater Cairo; the urban form is transitional at the moment leading perhaps to polycentric structure as structural growth lead by smart growth processes identify spatial disintegration and densification simultaneously thus concentrated distribution of heterogeneous land use patterns is manifest especially at the finite space of the Nile river valley and delta *References*	• Introverted urbanization and city primacy manifest with Dakar as the focal point; the urban form is somehow the case in Egypt tends towards unplanned polycentric structure indicated by urban fragmentation due mainly to dispersed land use distribution and characterized by large-grain heterogeneous urban pattern *References* UN-Habitat (2010: 116), Mbow et al. (2008: 76), Attahi et al. (2009), UN-Habitat (2010: 106), Kessides (2005: 70), World Bank (2002)	• Introverted urbanization and city primacy is characteristic in the context of quantitative and qualitative urban growth patterns; typically in Nairobi the urban form is experiencing spatially segregated dual urbanization occasioned by dispersed and concentrated distribution of large-grain homogeneous land use patterns which in turn are responsible for high density informal areas that accommodates the poor segment of the city and the low density wealthy residential neighbourhoods

(continued)

3.4 Basic Scenario of Planning in Select African Countries

Table 3.1 (continued)

Thematic areas	Sources: line of argument (country level)		
	Egypt	Senegal	Kenya
	Robson et al. (2012: 4, 5), World Bank (2007), Yin et al. (2005: 605), UN-Habitat (2010: 78), Ibrahim (1993: 132), Kessides (2005: 70)		*References* UN-Habitat (2005: 41, 2011: 5; 2008: 104, 115, 130; 2010: 139,141–145,153), Kessides (2005: 70)
Transboundary problems	• The incidence of sprawl, expansion and suburbanization is commonplace at the peri-urban areas leading to uneven spatial structures that supports extroverted urban economy	• The leapfrog model of urban sprawl coupled with suburbanization of population caused mainly by land market-driven functional rearrangement leads to the growth of informal settlements in peripheral areas	• Extroverted urban economy causing rapid growth of informal economy and corresponding development of informal settlements mostly at the fringe areas as poverty driven urbanization lead to urban expansion
	References World Bank (2007: 4–5), UN-Habitat (2010: 78, 2011: 2), Robson et al. (2012: 5), Wahdan (2007: 2099, 2101, 2102), Madbouly (2009: 76/81)	*References* World Bank (2002), Mbow et al. (2008: 76, 79), UN-Habitat (2010: 106, 110–116), GSDI (2010: 25)	*References* UN-Habitat (2005: 11, 16; 2010:168,140), Kabwegyere (1979: 311), GSDI (2010: 25)
Levels of spatial/economic development planning	• As in DRC planning Egypt is stratified along territorial levels: programme planning at local level, master planning at regional level, and strategic development framework at national level	• Stratified planning perspective is manifest: development planning at national level, strategic planning at regional level, and a combination of land use master planning and upgrading programmes at local level	• Development planning most likely resource based and sectoral is applied across board
	References Ibrahim (1993:127), Robson et al. (2012: 5), World Bank (2007: 6) Doc(s): Strategic Urban Development Plan (SUDP) (2008–2011)	*References* Mbow et al. (2008: 76), USAID/Senegal (2012: 13), Attahi et al. (2009: 18), UN-Habitat (2010: 121), World Bank (2002)	*References* Green (1965), Mabogunje (1990: 153), UN-Habitat (2010: 169), Kabwegyere (1979: 308), ETC (2006: iv) Doc(s): Integrated Regional Development Plans (IRDPs)

(continued)

Table 3.1 (continued)

Thematic areas	Sources: line of argument (country level)		
	Egypt	Senegal	Kenya
Urban planning instruments	• Transition to strategic planning using participatory process for planning initiative at all levels *References* Madbouly (2009: 81, 99)	• Development agencies induced commitment to participatory planning perspective at all levels of planning initiative *References* World Bank (2002)	• Master planning approach for regional and local planning initiative *References* Omwenga (2001: 1), UN-Habitat (2008: 122, 124)
Planning education facilities	• With three university facilities and a research institute in planning at Rabat planning education is receiving some attention; the curriculum of the university facilities are in the process of change under the influence of neoliberal planning *References* Madbouly (2009: 103)	• Content with the facility in Lome shared by all Francophone countries *References* Attahi et al. (2009: 51)	• Four planning schools at university level exist in Kenya but it is only University of Nairobi that has urban and regional planning focus *References* AAPS (2010: 8)
Professionalism (Urban Planning)	• No formal structures in place for professional planning practice	• No formal structures in place for managing professional planning practice	• Kenyan planners at least are organized although there is no indication that the country has a registration council that controls professional practice
National Urban Development Strategy (NUDS)	• Engages a combination of strategies that is likely to lead up to the development of polycentric urban region in the Nile river valley and the delta and create new frontiers for growth centres *References* UN-Habitat (2010: 81), Madbouly (2009: 76), UN-Habitat (2008: 64), del Rio Garcia Luis et al. (2010), Ibrahim (1993: 129), El-Ehwany and El-Laithy (No year: 26)	• Polycentric primate city region strategy with Greater Dakar as the focal point	• Multiple strategies without a clear direction *References* UN-Habitat (2010: 169), Otiso (2005: 127), Bubba and Lamba (1991: 58)

(continued)

3.4 Basic Scenario of Planning in Select African Countries

Table 3.1 (continued)

Thematic areas	Sources: line of argument (country level)		
	Egypt	Senegal	Kenya
Development cum transport corridors/transnational corridors	• Ecological issues challenge the frugality of developing the coastal region corridor and this leaves Egypt with option of checking for alternative corridors as a more sustainable strategy *References* UN-Habitat (2008: 28, 66–67; 2010:99), Thomas (2009)	• The corridors in Senegal especially the transnational corridors are more or less transport corridors and Senegal should endeavour to serve as a node rather than a transit terminal *References* UN-Habitat (2010: 131), Thomas (2009)	• An initiative similar to that of Singapore's growth triangle is required in Kenya's transboundary corridor with DRC and Uganda *References* UN-Habitat (2008: 29), UN-Habitat (2008: 29), Thomas (2009)
The informal sector	• Egypt seems to be experiencing a modern version of informal sector that accumulates wealth. Its more critical impact is on the informalization of settlement development which grew. In mid-1960s. This generated Hugh population of slum dwellers that presented serious planning and development problems *References* ILO (2002: 1, 31–33); del Rio Garcia Luis et al. (2010: 19), Madbouly (2009: 50), UN-Habitat (2010: 67)	• In spite of the relatively modern type of informal sector supported by institutionalist perception that surrounds its presence state attitude towards it remain severely repressive. However trends in informal settlement development continue as the sector generates squatter population that degrades to slum dwellers *References* Granström (2009: 25–26, 29, 35–36, 42), Mbow et al. (2008: 77, 86), World Bank (2002), Attahi et al. (2009: 17)	• Since 1990s Kenya is increasingly receptive to the survivalist informal sector that is manifest in its shores perhaps because of the growth of Jua kali industry—mainly home-based enterprises that operates small scale, locally and at a subsistence level. By implication Kenya has to be receptive to informal settlement development that normally accompanies the informal sector *References* Daniels (2010: 19), Xaba et al. (2002: 26), Wamuthenya (2010), Freeman (1975: 21), UN-Habitat (2005: 32), Republic of Kenya (2002: 14)

(continued)

Table 3.1 (continued)

Thematic areas	Sources: line of argument (country level)		
	Egypt	Senegal	Kenya
Paradigm shift in spatial and statutory planning	• Gone beyond transition from master planning and well within participatory planning practice in their planning history with the inception of Shorouk planning in mid-1990s. The activities of USAID, UNDP, UN-HABITAT, and GTZ are not disconnected with this speedy transition *References* del Rio Garcia Luis et al. (2010: 18), Madbouly (2009: 36/37, 48), El-Ehwany and El-Laithy (No year: 26); del Rio Garcia Luis et al. (2010: 22, 24)	• Partial shift from master planning paradigm as most of the attempts on participatory planning is still limited to slum clearance and urban upgrading programmes although a platform has been raised in 1996 for a determined consolidation of participatory planning practice *References* Attahi et al. (2009: 18, 34), World Bank (2002)	• Approaching two decades of donor-supported reforms to decentralize planning decision which was initiated in the 1990s is yet to yield any reasonable result. The strategy which is the hallmark of participatory planning simply does not resonate in practice with local institutions *References* UN-Habitat (2008: 124), Okpala (2009: 21), UN-Habitat (2005: 6–7,12; 2005b: 12, 65)
NEPAD	• Very active as a founding member of NEPAD. Facilitates most of the conferences *References* AMCEN (2008: 10), UN-Habitat (2005/6: 1)	• Fairly active member involved with many of NEPAD's networks *References* AMCEN (2008: 10), NEPAD (2002: 203)	• Shares similar participatory status with Senegal as a fairly active member but lacks initiative for NEPAD activities *References* AMCEN (2008: 10), UN-Habitat (2005/6: 1)
Urban planning consultancy services	• Both foreign and local expertise is active although national consultancy services are more active in planning for smaller towns that surround Greater Cairo *References* Ibrahim (1993: 124), del Rio Garcia Luis et al. (2010: 22)	• There is no indication that national consultants are trusted with the provision of consultancy services at best they are involved in foreign collaboration services *References* Attahi et al. (2009: 34), World Bank (2002)	• National consultants are not entrusted with the provision of consultancy services at best they work in partnership with foreign consultants *References* Oyugi and K'Akumu (2007: 101)

(continued)

3.4 Basic Scenario of Planning in Select African Countries

Table 3.1 (continued)

Thematic areas	Sources: line of argument (country level)		
	Nigeria	South Africa	Tanzania
Thematic areas	Sources: line of argument (country level)		
	Nigeria	South Africa	Tanzania
Country profile (growth indicators)	• Gained independence five decades ago, occupies tenth position in landmass and first position in population in the context of two-tier territorial divisions; more than half of the growing population is under poverty line and they engage mainly in agricultural activities • The national economy is driven by the industrial sector (precisely petrochemicals) and per capita productivity is recorded to be increasing while inequality is stabilizing at a mean position • The level of urbanization is increasing and megacity syndrome is evident • Actively involved as a regional giant in networking activities particularly for the West African region and the African region in general and hopes to the top-20 economies in the world *References* Olujimi (2009: 202), McGranahan et al. (2009: 43), UN-Habitat (2010: 63), Yousif (2005: 58), LOC (2008: 11–14), Kessides (2005: 71), Aigbokhan (2008: 14)	• Gained independence a century ago, sixth in landmass and seventh in population, operates four tiers of territorial divisions; the growing population engages dominantly in industrial (non-agricultural) activities and poverty structure is segregated fundamentally thus confirming huge inequality that is mainly responsible for the dysfunctional grading of the state • The national economy which is service driven is outstandingly the strongest in the region however with an impressive but deceptive per capita productivity and positive but unstable growth rate • The level of urbanization is increasing at not too high rate and prospects of megacities are high • Actively involved as a regional giant in networking generally for the African region but particularly for the Southern African region; committed to building a prosperous and egalitarian nation *References* Marx and Charlton (2003: 1), McGranahan et al. (2009: 43), UN-Habitat (2010: 207), Gelb (2002: 14), Akinboade and Lalthapersad-Pillay (2005: 244), van der Merwe (2008/9: 32), UN (2002: 42, 88), Kessides (2005: 71), Gelb (2003: 4, 5, 30), de Beer (2001: 1), UN-Habitat (2005/6: 4)	• Gained independence five decades ago with landmass just a little above that of Nigeria and almost at par with South Africa in population dynamics; the population engages dominantly in agricultural activities, about half of them are considered poor • Agriculture dominates the national economy which continually records high growth rate since the 1990s but per capita productivity stagnated in early 2000 although there are signs of phenomenal improvement since 2010 and inequality is low • Urbanization level is low and actually declined within the 2000–2010 periods and the primacy of Dar es Salaam remains constant • Involved in regional networking and hopes to develop an economy that will adapt to the changing market-space economy in the regional and global economy *References* Kironde, 2009:5, McGranahan, et al. 2009:43;UN-Habitat, 2010: 53,138; World Bank, 2002b; Kessides, 2005:71; Lupala & Nnkya, 2008:2; UN-Habitat, 2009:12; Abebe, 2011:2;UN-Habitat, 2009:12.

(continued)

Table 3.1 (continued)

Thematic areas	Sources: line of argument (country level)		
	Nigeria	South Africa	Tanzania
Urban planning machinery	• Amongst other related policies Nigeria has a defined urban development policy that is backed with urban and regional planning law enacted in early 1990s, the latest in the string of planning related legislation in the country; there are more provisional than there are substantive planning agencies and functions usually overlap sometimes leading to interdepartmental conflicts; meanwhile government adoptive bye-laws of the 1960s remain statutorily relevant but scarcely engaged in determining planning regulations nowadays except perhaps for prosecuting contraventions which is rarely the case • Summarily the planning machinery is equipped to function effectively *References* Okeke (2004: 195), AMCEN (2008: 12), Mabogunje (1990: 139, 141), UNCHS (1999:18, 25)	• South Africa is also engaged with decentralization policy and since the 1990s have endeavoured to backup this policy with appropriate legislation and development perspectives; the municipalities double as administrative and planning agencies hopefully they will liaise with the newly established National Planning Commission meanwhile the building bye-laws of the 1970s remains unchanged as it is the case in Nigeria and Kenya and this perhaps reflects their redundancy *References* Schoeman (2010: 23, 24), Diaw et al. (2002: 343), Lewis (2008: 8)	• Tanzania is engaged in decentralization policy but it does not exclude settlement development policies and their environmentally related policies and their planning laws are duly updated; remarkably and contrary to the situation in most African countries Tanzania has formal updated National Planning Standards although informal standards that govern planning practice derive from these national standards through participatory process and secondly the multiplicity of planning agencies does not seem to obtain *References* AMCEN (2008: 12); URT (2008: ix); Lupala and Nnkya (2008: 19), UN-Habitat (2010: 47), UNCHS (1999: 16, 20, 25, 35), URT (1997: 8)
Urban form	• Contrary to the general trend in Africa urbanization pattern in Nigeria is extroverted and this is not unrelated to the political administrative system of decentralizing governance; the urban form that responds to concentrated and dispersed distribution of land use	• Introverted urbanization spread amongst big cities is manifest but structural growth leading to urban change is taking place at the moment suggesting a transition most likely to extended metropolitan regions as home lands gradually assume the status of functional urban areas; the urban form	• Introverted urbanization dominates but extroverted tendencies are manifest in the urban hierarchy based on regional variations in urban perspectives; the urban form is reminiscent of the situation in Kenya where spatially segregated dual urbanism—represented by planned

(continued)

3.4 Basic Scenario of Planning in Select African Countries

Table 3.1 (continued)

Thematic areas	Sources: line of argument (country level)			
	Nigeria	South Africa		Tanzania
	development manifest spatially segregated high and low density heterogeneous urban patterns that are driven by qualitative and quantitative urban growth *References* UN-Habitat (2010: 116), Nwaka (2005), Onokerhoraye (1978: 174); UN-Habitat (2008: 95), Oduwaye (2009: 161)	manifest sprawl distribution of heterogeneous urban patterns that respond mainly to succession urban processes than gentrification which sparingly lead to densification *References* Küsel (2009: 2), Dauskardt (1993: 11), Watson (2008: 9, 10), Mammon (2005: 4), Cilliers (2010: 18), Donaldson (2001: 4, 8), Todes (2012: 159), Kessides (2005: 70)		concentrated distribution of land use and informal dispersed distribution responsible for urban fragmentation and transition from the heritage of large-grain homogeneous urban patterns to currently heterogeneous patterns especially in the expanding vertical slums—coexist as the urban area (specifically Dar es Salaam) experience quantitative and qualitative spatial growth *References* UN-Habitat (2008:151, 2010: 139,144, 145), Abebe (2011: 13), URT (2004: 77), Sawers (1989: 841/854), Hill and Linder (2010: 1/2), Kessides (2005: 70)
Transboundary problems	• Urban expansion mainly in the form of commercial ribbon street and peripheral slum development programmed to promote trade on foreign goods *References* Oduwaye and Lawanson (2008: 4), Agbola (2008: 13), Onibokun and Kumuyi (1996), Onokerhoraye (1978: 174, 176), Hicks (1998), Taylor (1988: 5), UN-Habitat (2010: 110–116); GSDI (2010: 25)	• Spatial fragmentation and incipient low density spatial expansion leading to edge cities and continual mushrooming of informal settlement on the urban edge *References* Mabin and Smit (1997: 207), Donaldson (2001: 8), Awuor-Hayangah (2008: 25), Mammon (2005: 4, 7), Orange (2008: 1, 8)		• Introverted but survivalist economy that encourages Greenfield development in the form of informal subcentres in peri-urban areas—as spatial form of Dar es Salaam increasingly assume ribbon-like leapfrog pattern of land use distribution *References* Hansen and Vaa (2004: 143), Abebe (2011: 16), Hill and Linder (2010: 1), UN-Habitat (2010: 13), URT (2004), Xaba et al. (2002: 5), GSDI (2010: 25)

(continued)

Table 3.1 (continued)

Thematic areas	Sources: line of argument (country level)		
	Nigeria	South Africa	Tanzania
Levels of spatial/economic development planning	• Hierarchical strategic planning initiative that adopts multiple planning frameworks ranging from broad guidelines to detailed master plan frameworks *References* Green (1965), Okeke (2004: 192), FGN (2001: 1) Doc(s): National Physical Development Plan (NPDP) (2010–2030); Niger Delta Regional Development Master Plan (NDRDMP) (2005–2020); State Economic Empowerment and Development Strategy (2005) (SEED); Local Economic Empowerment and Development Strategy (LEED), etc.	• Multiple categories of spatial planning frameworks heavily inclined towards the provision of broad guidelines for development *References* Van der Merwe (2008/9: 5, 8, 13), UN (2002: 7) Doc(s): National Spatial Development Perspective (NSDP)(2003); Provincial Spatial Development Strategy/Framework (PSDS); Provincial Growth and Development Strategy (PGDS). etc.	• Land use planning most likely broad based at national level leads strategic development planning at regional and local levels *References* Cooksey and Kikula (2005: 9), Armstrong (1986: 53), URT (2004: 29) Doc(s): National Land Use Framework Plan (2008–2028)
Urban planning instrument	• Admixture of design-oriented statutory master planning for city planning initiative and multiple planning frameworks derived through participatory process for resource planning and project development *References* Okeke (2004: 80)	• Adopts participatory process for spatial planning expressed in integrated development frameworks *References* Gordon and Lincoln (2008: 5). Harrison (No year: 186)	• Design-oriented statutory master planning allegedly transiting to non-statutory participatory planning approach to prepare EPM (Environmental Planning and Management) frameworks *References* Lupala and Nnkya (2008: 8, 19), Armstrong (1986), Hansen and Vaa (2004), Abebe (2011: 29)
Planning education facilities	• Nigeria has the highest number of planning schools at university level, in addition to polytechnics and a research institute.	• South Africa occupies the second position in the provision of university level planning education facilities including	• With two university facilities for planning education and an institute for rural development planning education Tanzania

(continued)

3.4 Basic Scenario of Planning in Select African Countries

Table 3.1 (continued)

Thematic areas	Sources: line of argument (country level)		
	Nigeria	South Africa	Tanzania
	Except perhaps for the research institute the curriculum of these institutions is design oriented based on multidisciplinary theoretical foundations to produce planners who are favourably disposed to multidisciplinary teamwork *References* Oduwaye and Lawanson (2008: 32)	technical colleges. Since 1980s many planning institutions have redeveloped their curriculum towards the American model of knowledge-based social science planning education while some others maintains fate with traditional physical planning approach *References* Diaw et al. (2002: 344), AAPS (2010: 9)	is not properly disposed for planning education. Meanwhile available planning schools contend with curriculum adjustments leading up to collaborative and participatory urban planning education *References* Odendaal (2012: 175–177)
Professionalism (urban planning)	• Nigeria has full organizational structures for managing professional planning practice	• South Africa has full organizational structures for managing professional planning practice *References* Tapela (2008: 20)	• Tanzania has full organizational structures for managing professional planning practice *References* Lupala and Nnkya (2008: 19)
National Urban Development Strategy (NUDS)	• Growth centres strategy linked with governance and the accretion growth of development corridors lead by market-space economy in which the growth centres serve as nodes	• Multiple strategies that have no clear direction although the spatial development initiatives (SDIs) and development corridors (DCs) tend to dominate the scene *References* UN-Habitat (2008: 1–25), Awuor-Hayangah (2008: 1)	• Another case of multiplicity of strategies and probably each strategy is not given time to mature before it is overturned *References* UN-Habitat (2008:124, 2010: 147), UNCHS (1999: 30), Mosha (2008: 51), Sawers (1989: 845)
Development cum transport corridors/transnational corridors	• Prospects of corridor development are high but the required planning consideration is lacking. Current accretion growth needs to be better coordinated with space economy strategies	• Development corridor strategy remains the cardinal strategy deployed for space economy development. Having applied this strategy since early 1990s the result is surprisingly not impressive given the	• Tanzania is one of those African countries with typically transport corridors that should be more strategic in planning their space economy to avoid being used as transit terminals rather than development

(continued)

Table 3.1 (continued)

Thematic areas	Sources: line of argument (country level)		
	Nigeria	South Africa	Tanzania
	References Taylor (1988: 5), UN-Habitat (2008: 114), Thomas (2009)	perceptions in literature. As presently constituted the strategy tend to disregard the development of homeland areas, marginally impact growth and yet unable to address the shape of urban regions. As it were its statutory status demands attention *References* UN-Habitat (2008: 28; 2010: 235(6), de Beer (2001: 6, 15), Thomas (2009)	node *References* Sawers (1989: 843), Thomas (2009)
The informal sector	• Attitudinal change from repression to pragmatic and promotional towards the informal sector in Nigeria is also manifest in the 1990s as it is the case in Kenya. Perhaps neoliberal thinking that erupted within the same period in Africa along with the institutionalist perception of the sector is responsible for these changes in attitude *References* Arimah (2001: 116), Nwaka (2005), Xaba et al. (2002: 9), Onyenechere (2011), Kessides (2005: 18), Oduwaye and Lawanson (2008: 8), Obinna et al. (2010)	• The support given to informal sector needs to be viewed against the backdrop of acute inequality in the society and the rationality of the support will be clearer *References* Awuor-Hayangah (2008: 12), Xaba et al. (2002: 12, 25), Davies and Thurlow (2009: 8, 13, 20), Awuor-Hayangah (2008: 7)	• Even before the boom period for informal sector proliferation in the 1990s Tanzania supported the sector. In fact as at the boom period Tanzania was moving ahead with the formalization of property rights in unplanned urban settlements in Dar es Salaam. Meanwhile the growth of informal settlements continues unabated without immediate solutions *References* Xaba et al. (2002: 5, 26, 27), Garcia-Bolivar (2006: 20), Malele (2009: 28), URT (2008: x)
Paradigm shift in spatial and statutory planning	• Master planning seems to remain resilient in practice in spite of overtures for alternative paradigms which at the moment remain mainly at the realm of rhetoric	• In the course of reforms in planning it seems paradigm shift is concerned more with the legal status and level of details expected in spatial planning frameworks.	• Statutorily master planning is still on board while participatory planning is marauding seeking for an entry point to secure committed acceptance. The usual leeway

(continued)

3.4 Basic Scenario of Planning in Select African Countries

Table 3.1 (continued)

Thematic areas	Sources: line of argument (country level)		
	Nigeria	South Africa	Tanzania
	except perhaps the national visioning process. All of the structures for master planning are still very much intact and planning decisions although very much prone to abuse by the political class is not liberalized *References* Okeke (1998: 134), Okpala (2009: 12–13), NTTP (No year: 57–59, 66, 105, 116)	At the moment South Africa is yet to strike a balance but reasonable stability has been achieved *References* Oranje (2002: 4, 6), Mabin and Smit (1997), Watson (2008: 9), Harrison (No year: 1)	of slum clearance and urban upgrading programmes used by donor agencies to instil participatory planning seem not to be securing the desired commitment to oust master planning *References* Mosha (2008: 51), Abebe (2011: 30), World Bank (2002b), Cooksey and Kikula (2005: 9), Lupala and Nnkya (2008: 7), UN-Habitat (2010b: 27), Hansen and Vaa (2004: 144), UN-Habitat (2006: 2)
NEPAD	• Active foundation member that motivates facilitates and funds NEPAD activities *References* UN (2004: 4, 23)	• Active foundation member that sustains NEPAD spirit provides leadership in charting the course for NEPAD activities and practically involved in determining NEPAD planning instruments *References* UN (2004: 3, 31, 33), OSISA and Oxfam (2009: 29), AMCEN (2008: 10), NEPAD (2012)	• Active but lacks initiative for NEPAD activities *References* AMCEN (2008: 10), UN-Habitat (2005/6: 1)
Urban planning consultancy services	• Most of the consultancy services are provided by foreign consultants although that was in the 1960–1980 periods. Nowadays partnership between foreign and national consultants is encouraged *References* Mabogunje (1990: 148), Ilesanmi (1998: 76), Okeke (1998)	• Consultancy services before the 1960s were provided by foreign consultants. Consultancy services were discouraged thereafter and planning mandates were given to numerous national committees *References* Mabin and Smit (1997: 203, 206, 209)	• Foreign consultants took charge of the master planning era. With the transition to framework planning university consultancy service is sought although it is not clear the nationality of members of the university consultancy unit *Reference* Armstrong (1986: 44/49), Lupala and Nnkya (2008: 7, 19)

(continued)

Table 3.1 (continued)

Thematic areas	Sources: line of argument (country level)
	Ethiopia
Thematic areas	Sources: line of argument (country level)
	Ethiopia
Country profile (growth indicators)	• Was never really colonized discounting short-lived Italian occupation in the 1970s, ninth in landmass and second in population, operates four tiers of territorial divisions; poverty situation is reportedly declining in the growing population who are mainly engaged in agricultural activities • As at year 2000 the national economy is terribly low but has reportedly grown phenomenally thereafter at consistently very high growth rates and inequality is low • The level of urbanization is low but increasing as it is the case in Tanzania and the primacy of Addis Ababa remains constant • The state is involved in regional networking activities and hopes to build cities that are centres of commercial and industrial development *References* Tolon (2008: 5, 6, 9), FDRE (2009: 2), McGranaham et al. (2009: 43), Yousif (2005: 58), UN-Habitat (2010: 53, 142), UN (2002b: 15)
Urban planning machinery	• As it is the case in Tanzania decentralization policy is operational along with urban development policy and other related environmental policies each properly backed up with legislation; from 2005 the mandate for urban planning transferred to a singular federal ministry presumably with subsidiary offices at different territorial divisions and by 2007 the planning agency was provided with operational planning standards subsumed in Regulation No. 135/2007 contrary to city-specific regulations of the 1980s *References* Tolon (2008: 23), Kedir and Schmidt (2009: 18), UN (2002b: 17, 26), FDRE (2006: 3, 4, 20), Asfaw et al. (2011: 19, 31)
Urban form	• Urbanization in Ethiopia is introverted and the urban system is dominated by the only big city, Addis Ababa; the urban form is dispersed comprising of heterogeneous urban pattern interspersed with nucleated homogeneous patterns that leads to gated communities; qualitative and quantitative growth and gentrification processes influence the dynamics of the urban form causing densification particularly in Addis Ababa *References* Asfaw et al. (2011: 2, 18, 34), UN-Habitat (2010: 150), Tolon (2008: 27/28, 30), UN (2002b: 9), Kessides (2005: 68)

(continued)

3.4 Basic Scenario of Planning in Select African Countries

Table 3.1 (continued)

Thematic areas	Sources: line of argument (country level)
	Ethiopia
Transboundary problems	• Poorly developed urban economic base and the incidence of squatter settlements leading to slum formation especially in the peripheral area of Addis Ababa
	References
	UN-Habitat (2010: 150), Melesse (2005: 11), Asfaw et al. (2011: 18); FDRE (2006: 16), UN-Habitat (2007), Tolon (2008: 28)
Levels of spatial/economic development planning	• Essentially engages in project planning for sectoral development
	References
	Kedir and Schmidt (2009: 13), FDRE (2006: 32), Tolon (2008: 19, 24), UN-Habitat (2008: 10), UN (2002b: 9, 13) Doc(s): 2002 **Addis Ababa** Structure Plan; Market Towns Development Project (MTDP), etc.
Urban planning instruments	• Participatory planning approach at all levels of planning initiative
Planning education facilities	• Considering the population of Ethiopia the provision of two university facilities for planning education plus an institute of technology and a college is considered grossly inadequate. In what looks like shared responsibility the university curriculum has a strong technical design/physical planning focus while the college concentrates on social sciences and humanities subjects
	References
	AAPS (2010: 7)
Professionalism (urban planning)	• No formal structures in place for managing professional planning practice
National Urban Development Strategy (NUDS)	• Inclined towards a polycentric model since the 1990s
	References
	UN-Habitat (2010: 137), Asfaw et al. (2011: 55)
Development cum transport corridors/transnational corridors	• The formation of national development corridors are most probably unfolding organically through natural process without formal planning that is required for securing productive space economy that generates growth
	References
	Kedir and Schmidt (2009: 10), Thomas (2009)

(continued)

Table 3.1 (continued)

Thematic areas	Sources: line of argument (country level)
	Ethiopia
The informal sector	• Institutionalist perception provides a theoretical foundation for the support of the informal sector. This is the case with Ethiopia where the informal sector makes significant contribution to the national economy however without relenting in generating informal settlements and squatter settlers
	References
	Pieter van Dijk and Fransen (2008), Tolon (2008: 29–30)
Paradigm shift in spatial and statutory planning	• Receptive to paradigm shifts at least to the extent of relaxing centralized planning and allowing some impetus for market forces to play some role in the process of rational land use planning which was introduced in the 1990s
	References
	Okpala (2009: 14), UN-Habitat (2008: 125), Tolon (2008)
NEPAD	• Active at networking in NEPAD activities
	References
	AMCEN (2008: 10)
Urban planning consultancy services	• The national planning institute is squarely in charge of providing consultancy services and they are busy with producing master plans
	References
	Doc(s): Ethiopian Master Plan

Source Own construction (2012)

3.4 Basic Scenario of Planning in Select African Countries

Table 3.2 Summary of perceptions on thematic areas studied in selected African countries

Thematic areas	Perceptions
Country profile (growth indicators)	• The Tanzanian and to some extent the Egyptian economy hold lessons for African countries because they were identified to have congruent indicators where the GDP composition correlates with employment structure • For most other countries the economic sector that powers the national economy consistently offers fewer employment opportunities and in extreme cases such as in Nigeria and South Africa inequality is high and poverty is growing because the majority of the people are unproductive and immaterial in the husbandry of the national economy • An enclave economy such as in Angola and Nigeria, that depends on resource mining, especially petrochemical resources, gives a wrong impression of healthy national economic development, whereas in real terms economic growth does not reduce poverty. The increases recorded for GDP per capita do not reflect in the dynamics of poverty
Urban planning machinery	• Decentralization and environmental policies tend to remove attention from urban planning and development policies in most African countries except perhaps in Nigeria, Ethiopia and possibly Tanzania where there is a defined urban policy and in the case of Tanzania human settlement development policy • The review of planning laws is seldom extended to adaptive bye-laws that deal with planning regulations. It seems those bye-laws are already submerged in crises of relevance • A multiplicity of sectoral planning agencies is identified, especially for housing and infrastructure development and this has the effect of generating incremental, piecemeal, and disjointed planning with little interdepartmental coordination. In the scenario urban planning agencies retire into informal development control without bothering with planning control
Urban form	• Introverted urbanization is commonplace in Africa except in the rare cases of Nigeria and Tanzania where it is extroverted • Urban growth comprehension tends to be synonymous with demographic increases, in total disregard of the very dynamic spatial aspects of it that determine urban form • Urban form is the least issue planning practice is concerned with in Africa; the same neglect reflects in planning education with the result that elements of urban form that should inform planning education have paled into insignificance and in the process

(continued)

Table 3.2 (continued)

Thematic areas	Perceptions
	enhanced the stakes of generic procedural planning education
	• There is a relative lack of knowledge of the form of African cities expressed in terms of the attributes of key variables of urban form, theoretically identified in the interplay of urban systems (urbanization patterns), urban land use distribution and urban land use patterns. What is available is implied knowledge and not results of research or monitoring activities
Transboundary problems	• Sprawl urban expansion and suburbanization are evident and are identified as being led by the proliferation of informal settlement development
	• Extroverted national economies are overwhelmed by the development of survivalist informal sector operations
	• Dualistic and fragmented space economy stratified spatially along the lines of rural and urban economies and resulting in uneven spatial structures (distortions) in the urban regions
Levels of spatial/economic development planning	• Broad guideline frameworks determined by market forces are extending into regional and local planning thus creating a lacuna in framework planning that deals with shaping the city
Urban planning systems	• In practical terms paradigm shifts are not succeeding in reinventing planning
	• Design-oriented master planning perspective is remarkably resilient
	• A compromise position in planning perspective is settling on new master planning which incorporates participatory process although it is yet unclear the best way to find a synthesis between the two entities of participatory process and planning process
	• Transitions in planning systems in most African countries started in the 1990s when neoliberal economic policies took charge of global economy
	• The transition in most countries is either induced by donor agencies or lead by conformity to global trend which the political class find strategically expedient for their countries to remain relevant in global political economy and not necessarily in the best interest of developing their urban regions
Planning education facilities	• Francophone countries seem to be deficient in the provision of planning education facilities
	• Neoliberal thinking is influencing changes in the curriculum of planning education. Most of the changes are focused on generic procedural planning although progress in this direction is stunted

(continued)

3.4 Basic Scenario of Planning in Select African Countries

Table 3.2 (continued)

Thematic areas	Perceptions
	• Knowledge production in substantive planning theories is deficient in Africa and unfortunately against the backdrop of the need to review the Universalist approach to planning theory
Professionalism (urban planning)	• The organization of professional planning in Africa is abysmally poor, not including here the exceptional situation in Nigeria, South Africa and Tanzania
National Urban Development Strategy (NUDS)	• Urban region development for metropolitan areas of mostly primate cities is commonplace. How this correlates with the decentralization policy of most countries is not clear especially where they are not related to any known urban region development model • Modelling of urban development seems to be caught in the web of informality; however, this has implications for urban productivity because not all urban forms generate growth
Development cum transport corridors/transnational corridors	• So-called development corridors if not properly conceived and managed may indeed turn out to be conduits for external economies to infiltrate the space economy of African countries • Ensuring that development nodes in the corridors are not transit terminals as it seems to be the case with Maputo in Mozambique is critical • At the moment market forces serve as determinant factors for development corridors in Africa and it is not clear how this will facilitate regional integration in Africa
The informal sector	• The informalization of human systems tends to paralysis the planning system • Attitudes towards the informal economic sector are fast moving from repressive to optimistic although there are rare cases such as in Mali and Tanzania where hostile regulatory environments remain resilient • The change of attitude towards the sector is without clear direction of how to deal with the sector's externalities on spatial development of the urban region • Reformist perceptions, as is the case in most African countries, give a more realistic perspective on the growth of survivalist informal sectors in Africa
Paradigm shift in spatial and statutory planning	• Within three decades of crusade for participatory planning the performance sheet indicates limited progress as typified in Egypt under pressure from external development partners. On the other hand stalemated situations are commonplace while in limited instances such as in DRC and to some extent Tanzania zero tolerance is noticed

(continued)

Table 3.2 (continued)

Thematic areas	Perceptions
	• The entry point for instituting paradigm shift in planning in Africa is through slum clearance and urban upgrading schemes sponsored by donor agencies • Where attempts are made to apply the concept at the urban level, such as in South Africa, the tactic of non-statutory, broad guideline spatial planning is engaged and participation is more or less consultative
NEPAD	• South Africa is the prime mover of the NEPAD initiative since its adoption in 2001 • Most other countries are less enthusiastic and seem to be more concerned with what they can get out of the initiative in terms of attracting FDIs • NEPAD does not have the planning tools to deal with regional integration. The RAIDS strategy is a move towards this direction but there are lots of matters arising with the strategy especially the strategic implications of sitting of the proposed development corridors at the edge of the continent and achieving form-based growth
Urban planning consultancy services	• Planning consultancy services in Africa is extroverted in favour of foreign planning consultants. But this is to be expected because many African countries do not have organized professional practice in planning

Source Own construction (2012)

Given the summary of perceptions, the planning scene manifests some tendencies. The object of planning is poverty reduction, which does not necessarily imply to enhance productivity. This explains the receptive attitude towards the informal sector. Incidentally, the informal sector that is manifest in Africa is that of the survivalist category, which may enhance survival but retains poverty. The survivalist informal sector is built on technocentrism. The technocentric mode of modern environmentalism perceives the natural environment as a natural stuff from which man can profitably shape his destiny. Because of the environment problems associated with this mode of use of the environment, attention tend to drift from urban planning to environmental management, which is more or less a cross-cutting issue in urban planning. The disregard of the urban environment as the subject of spatial planning exacerbates the urban crisis. Ultimately, priority transnational problems manifest unstable urban forms, dysfunctional activity systems in the urban regions, slum formation, isolated urban hierarchies, and sprawl development. These problems are left unattended because the planning system, which is progressively under the influence of market force, is not inclined towards spatial integration. The emerging planning initiatives sponsored mainly by UN-Habitat subsidiaries are more inclined towards project development.

3.5 Conclusion

The resilience of epistemological foundations and ideologies, which accounts for imperial space economy and informalization, is also responsible for the gross misreading of the African crisis in the world system (Okeke et al. 2016). It seems the African crisis is defined to be synonymous with the externalities of neoliberal economy and this mindset informs the planning scene. Informality in planning is the vogue and the informalization of cities is underway in line with neoliberal philosophy. Efforts to reverse this trend are torpid.

The re-engineering of the epistemological foundations and ideologies identified has long been identified as the entry point to work Africa back into the mainstream of capitalism. The problem with engaging this task lies with the oversight of meta-theoretical matters in planning theory. The generators of these epistemologies reside in development ideologies found the meta-theoretical realm. Planning responds to development ideologies and development ideologies have appropriate planning response. Hence for neoliberal development ideology, neoliberal planning theory suffices.

References

Albers G (1986) Changes in German town planning: a review of the last sixty years. Town Plann Rev 57(1):22

Amin S (ed) (1974) Modern migrations in West Africa. Oxford University Press, London

Beinfield MA (1975) The informal sector and peripheral capitalism: the case of Tanzania. IDS Bull 6(3):53–73

Bibangambah JR (1992) Macro-level constraints and the growth of the informal sector in Uganda. In: Becker J, Pedersen P (eds) The rural-urban interface in Africa, Uppsala. The Scandinavian Institute for African Studies

Breman J (1999) Industrial Labour in Post-Colonial India. II: employment in the informal-sector economy. Int Rev Soc Hist 44:451–483

Cardoso FH, Enzo F (1969) Dependencies of Desarrollo en America Latina. Siglo XXI, Mexico DF

Castells M, Portes A (1989) World underneath: the origins, dynamics, and effects of the informal economy. In: Portes A, Castells M, Menton LA (eds) The informal economy: studies in advanced and less developed countries. John Hopkins University Press, Baltimore, pp 11–40

Christaller W (1933) 1966. Central Places in Southern Germany (trans: Baskin CW). Prentice-Hall, Englewood Cliffs, NJ

Chukuezi CO (2010) Urban informal sector and unemployment in Third World cities: the situation in Nigeria. Asian Soc Sci 6(8):133p

Clammer J (1975) Economic anthropology and the sociology of development. In: Oxaal I, Barnett T, Booth D (eds) Beyond the sociology of development: economy and society in Latin America and Africa. Routledge and Kegan Paul, London

Davidson B (1978) Discovering African's past. Longman, London, pp 87–111

de Soto H (2000) The mystery of capital: Why capitalism triumphs in the West and fails everywhere else. Basic Books, New York

de-Soto H (1989) The other path: the invisible revolution in the Third World. Harper and Row, New York

Duminy J (2011) Literature survey: informality and planning. Being literature survey complied for the Urban Policies Programme of the policy-research network Women in informal Employment: Globalizing and Organizing (WIEGO)

Edwardo EE (1990) Community design and the culture of cities. Cambridge University Press, Cambridge, p 356p

Gallacher R (n.d.) Participatory planning processes. RALA Report No. 200. Food and Agriculture Organization, Rome, Italy

Guglar J, Flanagan WG (1978) Urbanization and social change in West Africa. Cambridge University Press, Cambridge

Hall P, Pfeiffer U (2000) Urban fiiture 21: a global agenda for 21st century cities. E & PN Spon, London

Healey P (1997) The revival of strategic spatial planning in Europe. In: Healey P, Khakee A, Motte A, Needham B (eds) Making strategic spatial plans: innovation in Europe. UCL Press, London, pp 3–19

Healey P (2004) The treatment of space and place in the new strategic spatial planning in Europe. Int J Urban Reg Res 28(1):45–67

Hicks J (1998) Enhancing the productivity of urban Africa. In: Proceedings of an international conference research community for the habitat agenda. Linking research and policy for the sustainability of human settlement held in Geneva, 6–8 July

Ishengoma E, Kappel R (2005) Formalization of informal enterprises: economic growth and poverty. Eschborn 10:12, 26 (GTZ, http://www.gtz.de)

Krueckeberg D (1995) The difficult character of property: to whom do things belong}. J Am Plann Assoc 5(3):301–309

Landman K (2004) Gated communities in South Africa: the challenge for spatial planning and land use management. Town Plann Rev 75(2):163

Lowder S (1986) Inside Third World cities. Chrome Helm, London

Martinez-Vela CA (2001) World systems theory. ESD 83

Meillassoux C (1972) From reproduction to production. Econ Soc 1(1):93–105

North DC (1956) Exports and regional economic growth: a reply. J Polit Econ 64(2):165–168

Nwaka GI (2005) The urban informal sector in Nigeria: towards economic development, environmental health, and social harmony. Global Urban Devel Mag 1(1)

Okeke DC (2015) An analysis of spatial development paradigm for enhancing regional integration within national and its supporting spatial systems in Africa. Available at:http://dspace.nwu.ac.za/bitstream/handle/10394/15485/Okeke_DC.pdf . Date of access 19th June 2016

Okeke DC, Cilliers EJ, Schoeman, CB (2016) Neo-mercantilism as development ideology—a conceptual approach to rethink the space economy in Africa. African Studies (Accepted for publication)

Oliver P (ed) (1971) Shelter in Africa. Barrie & Jenkins, London

Onyenechere EC (2011) The informal sector and the environment in Nigerian towns: what we know and what we still need to know. Res J Environ Earth Sci 3(1):63p

Owusu F (2007) Conceptualizing livelihood strategies in African cities: planning and development implications of multiple livelihood strategies. J Plann Educ Res 26(4):450–463

Pahl-Weber E, Henckel D, Klinge W, Gastprofessorin P, Schwarm DZ, Rutenik B, Besecke A (2009) Promoting spatial development by creating common mindscapes. COMMIN: the Baltic Spatial Conceptshare, pp 3–21

Parthasarathy D (2009) Rethinking urban informality: global flows and the time-spaces of religion and politics. Paper for presentation at the international conference on Urban Aspirations in Global Cities, 9–12 August 2009, Max Planck Institute for the Study of Religious and Ethnic Diversity, Gottingen, Germany

Portes A (1983) The informal sector: definition, controversy, and relation to national development. Review (Fernand Braudel Center) 7(1):156, 159, 171

Portes A, Walton J (1981) Labor, class and the international system. Academic Press, New York

Rabinowitz P (2013) Participatory approaches to planning community interventions. Available at: http://ctb.ku.edu/Date of access 15 Oct 2013

References

Rakodi C (1997) Global forces, urban change, and urban management in Africa. In: Carole R (ed) The urban challenge in Africa: growth and management of its large cities. United Nations University Press. Tokyo

Rakowski CA (1994) Convergence and divergence in the informal sector debate: a focus on Latin America 1984–92. World Dev 22(4):501–516

Rimsha A (1976) Town planning in hot climates. Mir Publishers, Moscow

Roberts B (1978) Of peasants. Edward Arnold, London

Roy A (2005) Urban informality: towards and epistemology of planning. J Am Plann Assoc 71(2):147–158

Roy A (2009) Why India cannot plan its cities: informality, insurgence and the idiom of urbanization. Plann Theory 8(1):76–87

Roy A, AlSayyad N (eds) (2004) Urban informality: transnational perspectives from the Middle East, Latin America, and South Asia. Lexington Books, Lanham

Santos N (1979) The shared space. Methuen, London

Sassen S (1990) The mobility of labour and capital: a study in international investment and labour flow. Cambridge University Press, Cambridge

Schrank A (2009) Understanding Latin American political economy: varieties of capitalism or fiscal sociology? Econ Soc 1:53–61

Simone A (1998) Urban processes and change in Africa. CODESRIA working papers 3/97 Senegal, p 12

Tiebout CM (1956) Exports and regional economic growth. J Polit Econ 64(2):160–64

Todes A, Karam A, Klug N, Malaza N (2010) Beyond master planning? New approaches to spatial planning in Ekurhuleni, South Africa. Habitat Int 34:414–416, 419

Todes A (2012) Urban growth and strategic spatial planning in Johannesburg, South Africa. Cities 29:158–165

UN-Habitat (2009) Sustainable urbanization: revisiting the role of urban planning, global report on human settlement. UN-Habitat, Nairobi

Wallerstein I (2004) World-systems analysis: an introduction. Duke University Press, USA, p 128p

Watson V (2009) Seeing from the South: refocusing urban planning on the globe's central urban issues. Urban Stud 46(11):2259–2275

Wolpe H (1975) The theory of internal colonialism: the South African Case. In: Oxaal I, Barnett T, Booth D (eds) Beyond the sociology of development. Routledge and Kegan Paul, London, pp 229–252

World BANK (1989) Sub-Saharan Africa: from crisis to sustainable growth. The World Bank, Washington, DC

Chapter 4
Meta-Theoretical Frameworks of (Spatial) Planning

Abstract There are traditional and international perspectives of meta-theories of planning. The traditional perspectives are embedded in worldviews and philosophies, which mark traditional communities as unique societies. This work reviewed *ujamaa* philosophy based on socialist principles in East Africa, the *ubuntu* ideology based on survivalist structures (i.e. the extended family) in Southern Africa and the *omenani* philosophy of Ibo socialism built on spirituality amongst the Ibos in South-Eastern Nigeria, West Africa. International perspectives portray meta-theories as trade methodologies. Classical liberalism nurtured popular design tradition in planning, fettered capitalism nurtured formal master planning theory and neoliberalism nurtured neoliberal participatory planning theory. The precepts of neoliberal planning theory amount to the reinvention of planning.

Keywords Ujamaa · Ubuntu · Omenani · Meta-theory · Socialism · Communism · Liberalism

4.1 Traditional Perspective

Most traditional communities in Africa have their way of life expressed in their worldview and religion, which mark them as unique societies. As there are many societies, so there are many philosophies. The *ujamaa* philosophy based on socialist principles in East Africa, the *ubuntu* ideology based on survivalist structures (i.e. the extended family) in Southern Africa and the *omenani* philosophy built on spirituality amongst the Ibos in South-Eastern Nigeria, West Africa are notable examples. It is not clear what transpired in North Africa. The *shorouk* idea representing 'sunrise' in Arabic which is embedded in community development initiatives is noted. But it is a relatively new phenomenon, which was initiated in 1994 as development strategy that relies on local leadership, youths and women as well as governmental assistance to development (El Mahdi 2002, p. 26). In this discourse attention is focused on the philosophies in sub-Saharan Africa.

Omenani is the philosophy of Ibo socialism, which is tied-up with their cosmology therefore their eco-centric use of environment. *Omenani* is yet another sermon on morality and spirituality in which transcendentalism is the code of conduct for life and the foundation for value systems. '*Omenan*' determined settlement patterns in a manner that is reminiscent of the garden city Ebenezer Howard theorized in 1925. *Omenani* permits two categories of settlements: the home towns and market towns. Spatially, market towns located at the interface of neighbouring home towns. This spatial definition informs the space economy within which the functional flow of trade relations is organized in time and space by local institutions to avoid overlap. The sovereignty of the home towns and its sustained focus on transcendental ends is protected as it performs its resource management functions.

The *ujamaa* ideology in East Africa traditionally is a mindset determined by socialist values. Essentially, it describes African socialism symbolized with the extended family system in much the same way the Industrial Revolution and class distinction founded European socialism. The Tanzanian government translated *ujamaa* into a political-economic management model in 1967 when it became a development ideology for the delivery of African socialism. Naturally, the Tanzanian *ujamaa* model is opposed to capitalism, which seeks to build a happy society on the basis of the exploitation of man by man; and it is equally opposed to doctrinaire socialism which seeks to build its happy society on a philosophy of inevitable conflict between man and man (Nyerere 1962). Africa socialism, which *ujamaa* represents is a traditional heritage. After being modelled as a management tool, some argue that *ujamaa* is not just a development theory but an ideology for the reconstruction of an imaginary relationship of individuals at the level of the state (Cornelli 2012, p. 24).

Ujamaa as a concept tilts heavily towards socialism, brotherhood and family hood. The core assumption of the concept is human equality and its primary principles (practices) are: love, classless society, everybody is a worker, and wealth is shared. Thus, *ujamaa* is an expression of human equality, popular democracy, state ownership of property, self-reliance and freedom (Cornelli 2012, p. 51). It is inferred from the literature that these principles, which informed the *ujamaa* concept, are reactions to slave trade, colonialism, disappearance of African institutional system, development of individualism and selfishness, poverty, dependence, etc.

Ujamaa is an ideology for the following reasons: it represents the imaginary relationship of individuals to their real condition of existence, it had functional necessity, it is required to transform the lives of the people, it had material existence, and it interpolates individuals as subjects (Cornelli 2012, p. 161). As an ideology, *ujamaa* is expected to perform integrative functions for a society that is segregated along tribal, religious and race lines. Also transformative functions are anticipated to minimize class domination. Apparently, it was not recognized that *ujamaa* had planning functions to perform. This function is probably one of the most critical functions of a development ideology that seldom receives attention in the African context. But planning concretizes a development ideology in space. The villagization program of *ujamaa* ideology and the coercion of people into the

4.1 Traditional Perspective

villages suggest the oversight of an *ujamaa* planning theory as basic requirement to consolidate the ideology in space.

The liquidation of *ujamaa* villages at the approach of neoliberalism remarkably expressed the collapse of *ujamaa* ideology. It is not for fun that neoliberalism approached with neoliberal planning theory. By mid-1980, following the change in leadership, Tanzania had embraced neoliberalism and signed agreement with the World Bank and IMF for structural adjustment program (SAP). Tanzania moved from socialist to capitalist state and adopted participatory planning by induction. It is deduced that socialism prompted by *Ujamaa* was a victim of systemic undermining and this is linked with Tanzanian government reliance on external funding. The disposition of Tanzanian government towards external funding provided capitalism the leeway to fight Tanzanian socialism. This fight was done with financial mechanisms. Apparently, *ujamaa* collapsed but its tenets are remarkably resilient and are still applied as yardstick for evaluating policies.

Ubuntu, on the other hand, expresses African communalism based on approved code of conduct, almost similar to the *omenani* of the Ibo race although *omenani* is emphatic on spirituality while *ubuntu* emphasizes common good and politics. The focus of *ubuntu* is understood for a society that is exposed to socialization into racism. *Ubuntu* philosophy holds that racism is socially constructed; it is not innate. Hence, *ubuntu* principles rest on spiritual rearmament, selflessness, and honesty, etc., which are intended to make the people believe in themselves. It presupposes people with a sense of trounce but determined to survive collectively. The impression is perceived that *ubuntu* is a defence mechanism, where strength is found in cohesion and hospitality. This attribute is at the background of the quality of being human which *ubuntu* preaches. *Ubuntu* presents the doctrine of owing our selfhood to others.

The stimulants for *ubuntu* limit the *ubuntu* ideology to character moulding and acting as moral boosting therapy. Its links and relations with growing the economy and the use of space are not clear. This is why its status as a development ideology is called to question. It is difficult for instance to relate *ubuntu* to neoliberalism. Perhaps it could be said that it is opposed to private profitability preached by neoliberalism. But this is only a presupposition because there are no explicit tenets of *ubuntu* that opposes private profitability. Moreover, *ubuntu* neither harbours any planning philosophy nor is it known to impact settlement development patterns. Its influence in considering development ideology for Africa in the context of 21st century global economy tends to be inevitably limited. But it can inform the outlook of neo-mercantile development ideology as a process of socialization most likely to produce the kind of selfless leadership required for change in African development.

4.2 International Perspective

The foremost international development ideology worldwide is arguably mercantilism. As a trade methodology, which emerged in the fifteenth century, mercantilism has very strong historical antecedence in the economic growth of wide raging civilizations. Along the line, it diffused with worldviews of earlier epoch, which cradled traditional civilizations, to grow modern economies. Mercantilism is therefore not a novelty in the development of national economies, neither indeed is its new outlook neo-mercantilism in global regionalism. Both concepts are usually not publicly acknowledged by public policy makers because the concepts are link with protectionism, which negates trade liberalization and by implication globalization (Raza 2007, p. 1). Ironically, in practical terms, they are both functional and currently responsible for bilateral trade relation issues that are rocking developed countries. Sometime in the 1980s, mercantilism lost favour to neoliberalism. By the turn of the century, the economic failures recorded in the context of neoliberalism tended to redirect attention subtly to mercantilism but in the preferred mould of 'neo-mercantilism' as de facto development ideology for the twenty-first century.

In this subsection, as an entry point to address the dynamics identified in mercantilism, the tenets of mercantilism and its application in Africa, including its impacts, will be examined retrospectively. Thereafter, the rationality of ideological change that tended towards neoliberalism and its impact on Africa will be shared. The incidence of Euro-American mercantilism marked with the instrumentality of neoliberal planning and legal reforms such as Africa Growth and Opportunity Act (AGOA) on trade relations are examined. Since the turn of the century, ideological change to neo-mercantilism is the vogue in the trade policies of the international community. The theoretical framework of this change will be examined and relationships established.

The meaning of the term 'mercantilism' is embroiled in long debates concerning how economic historians and social scientists perceive it (Coleman 1969). Economic historians apply discrete, time bound definitions while social scientists prefer dynamic definitions that are time dependent, which identify the salient features of the term as it spans through historic systems. Both approaches are relevant in planning studies and they apply as fortuitous circumstance permits. Mercantilism is the political-economic philosophy of sixteenth through eighteenth centuries characterized by the desire of nations to enrich themselves through the control of trade. It emerged as a system for managing economic growth through international trade as feudalism became incapable of regulating the new methods of production and distribution. It was a form of merchant capitalism relying on protectionism. Hence, the concept of mercantilism covers protectionist policies to promote national economic development (Hettne 2014, p. 210). European nation-states used it to enrich their own countries by encouraging exports and limiting imports. This implied the extension of market-space economy.

The term 'mercantilism' denotes the principles of the mercantilist system, sometimes understood as the identification of wealth with money: but more generally, the belief that the economic welfare of the state can only be secured by government regulation of a nationalist character (Hettne 2014). Therefore, the mercantilist logic is fundamentally political. The point is not necessarily the maximization of economic development but the optimization of political control, i.e. optimization in the sense that the marginal utility of administrative controls must not fall below zero (Hettne 2014, p. 207). The protection of the state, as far as economic processes are concerned, is the essence of mercantilism. So mercantilism is concerned with statism and stateness, where the former refers to the instrumental strength of the state and the later the integrative strength of the nation-state thus signalling distributional justice.

There are different traditions of mercantilism as there are national political economies in Europe. Very few European countries actually developed in accordance with the way the World Bank and IMF now recommend the underdeveloped countries to develop. Rather they followed, in varying degree, the old Listian formula, i.e. they were essentially mercantilists (Hettne 2014, p. 212). Africa was at the receiving end of European mercantilism. The European mercantilist operations reworked the space economy and created capitalism in Africa in the first place. The resultant imperial space economy ushered colonization, which administered twentieth century European mercantilism. Africa was not exposed to European mercantilism only; the continent was also exposed to Arab mercantilism. But European mercantilism was peculiar because it sought for controls in trade relations unlike Arab mercantilism, which simply focused on fair trade although Arab mercantilism had expansionist religious objectives. Nevertheless, the imperial system entrenched with several decades of European mercantilism ultimately extroverted the economy of African countries.

Classic mercantilism lost favour when the rising bourgeoisie grew tired of the limitations mercantilism placed on their actions. Liberating themselves required them to secure the policy of limited state action, which was termed 'laissez-faire'. In the 1970s the bourgeoisie turned against post-war social Keynesianism because, as with mercantilism, it was limiting their actions and they embraced the ideas of Milton Friedman. Following Friedman's Monetarist ideas, state action remained but very much in the background skewing the market in favour of the wealthy (Anarcho 2005). The resultant predictable and predicted massive economic collapse, which empowered the wealthy, caused social Keynesianism to be replaced with neoliberalism as the de facto state principle. Neoliberalism represented a combination of "laissez-faire" and military Keynesianism. Its rational is therefore not pro-poor as its practice is projected to portray. Neoliberalism essentially introduced another form of protectionism that is linked with the so-called NAIRU theory (Anarcho 2005). The Non-accelerating inflation rate of unemployment (NAIRU) theory simply means the unemployment rate below which the inflation rate begins to rise. The NAIRU was used to justify economic policies designed to weaken the power of labour and strengthen that of the bosses (Anarcho 2005). Therefore, trade relations are still premised on the wealth accumulation principles of mercantilism and state intervention redefined to reflect the focus on private profitability. The primary agenda is to establish a new 'post-Fordist' accumulation dynamic, which has links with globalization in agriculture (Mark and Potter 2004, p. 1). In other words, agriculture enjoys relatively higher measure of protection regardless the hegemony of neoliberal principles. With this operational principle, weak states are vulnerable to dependency and marginalization.

Since the turn of the century, mercantilism remains resilient in neoliberal context. This has had serious adverse impacts on African economy. The space economy manifests deepening distortions and urbanization trends manifest introverted patterns which are resolved as epistemologies of European mercantilism. The outcome of urban primacy is commonplace in which the modern economic sector disconnects with the local economy (see Hicks 1998). Lately the in-formalization of African cities is manifest as induced population mobility generates informal labour force into the urban area. This is largely responsible for the in-formalization of human systems. The population dynamics accounts for rapid urbanization, which provides the rationale to jettison formal planning vis-à-vis master planning. The alternative planning procedure, which neoliberal planning theory provides with its doctrine of market-oriented planning, consolidates the in-formalization process. Meanwhile, the receptive attitude towards the informal sector provides the workforce for marketing foreign goods in Africa. In essence, consumer economy is continuing while strategic efforts are underway in trade relations to sustain the informal sector and deliver twenty-first century Euro-American mercantilism in Africa. The African scholars who are receptive attitude to the informal sector do not seem to share this insight.

The mindset of policy makers is framed to accept that the informal sector will always exist (Gerxhani 2004: 294). But the existence of the sector is not nearly as important as to how it exists under the combined influence of endogenous and exogenous policy frameworks contained for instance in trade liberalization Africa Growth and Opportunity Act (AGOA), and wage subsidies. In a study of formal-informal economy linkages and employment in South Africa, Davies and Thurlow (2009) established inter alia that trade liberalization reduces national employment, increases formal employment, hurts informal producers and favours informal traders who benefit from lower import prices. In other words trade liberalization encourages traders rather than producers. The study also indicates that wage subsidies on low-skilled formal workers increase national employment, but hurts informal producers by heightening competition in domestic produce market. The tendency is for the modern informal sector to go down while the unproductive traditional (survivalist) informal sector remain and probably increase.

The AGOA concept, which derives from trade liberalization, purports to provide a reverse trend. In the first half of 2001 as American imports grew 7 % and exports jumped 35 % American industry reported new sourcing contracts with African suppliers and many inquiries from African firms seeking joint ventures or US expertise and inputs. But does this vindicate the intent of AGOA to grow African economy through robust trade relations with American companies. The answer is probably 'not'. Meanwhile as wage subsidies create attraction for elusive formal jobs and in the process generate exodus of job seekers to urban areas where they inevitably engage in informal retail trade activities.

In the past three decades, the incidence of the so-called New Regionalism seen as extended nationalism altered the scale of space for mercantilism (see Seers 1983 cited in Hettne 2014, p. 223). Mercantilism seized to act on national space and extended to act on global space. Hence, the prefix 'neo' was added to mercantilism due to change in emphasis from classical mercantilism on military development, to economic development, and its acceptance of a greater level of market determination of prices internally than was true of classical mercantilism. Neo-mercantilism is founded on the control of capital movement and discouraging of domestic consumption as a means of increasing foreign reserves and promoting capital development. This involves protectionism on a host of levels: both protection of domestic producers, discouraging of consumer imports, structural barriers to prevent entry of foreign companies into domestic markets, manipulation of the currency value against foreign currencies and limitations on foreign ownership of domestic corporations. The purpose is to develop export markets to developed countries, and selectively acquire strategic capital, while keeping ownership of the asset base in domestic hands. It therefore suggests new form of protectionism that is qualitatively different from the traditional mercantilist concern with state building and national power, i.e. the pursuit of statism.

Neo-mercantilism is a transnational phenomenon. It transcends the nation-state logic in arguing for a segmented world system, consisting of self-sufficient blocs. With this outlook, it deals with the modern industrial world as opposed to the pre-Industrial Revolution version of mercantilism. Its protagonists do not believe in the viability of closed national economies in the present stage of the development of the world economy. Therefore, neo-mercantilism as a concept is seen as a policy regime that encourages exports, discourages imports, control capital movement, and centralizes currency decisions in the hands of a central government. This means the pursuit of economic policies and institutional arrangements, which see net external surpluses as a crucial source of profits. Furthermore, neo-mercantilism is an economic theory that maximizes the benefits to and interests of a country such as higher prices for goods traded abroad, price stability, stability of supply and expansion of exports with concomitant reduction of imports.

Neo-mercantilists believe in the regionalization of the world into more or less self-sufficient blocs where political stability and social welfare are major concerns. Therefore, the aim and objectives of neo-mercantilist policies is to increase the level of foreign reserves held by government; it allows more effective monetary policy and fiscal policy. This is generally believed to come at the cost of lower standards of living than an open economy would bring at the same time, but offers the advantages to the government in question of having greater autonomy and control. Its policy recommendations are generally protectionist measures in the form of high tariffs and other import restrictions to protect domestic industries combined with government intervention to promote industrial growth, especially manufacturing. China, Japan and Singapore are described as neo-mercantilist.

Market economists have argued the pros and cons of protectionism. Nevertheless, the language of neo-mercantilist policies repeats the claims of earlier centuries that protective measures benefit the nation as a whole and that governmental intervention secures the 'wealth of the nation' for future generations. Indeed, the historical evidence leads any unbiased researcher to conclude that mercantilism has generally been successful in growing economic development. Free-trade advocates have failed to muster counter-arguments for why Britain fell behind the United States and Germany by 1880 after having abandoned mercantilism in favour of free-trade in the middle nineteenth century.

4.3 Neoliberalism as Meta-Theory of Planning

The historical antecedence of neoliberalism has the 1930s, 1960s and 1980s as critical turning points. Following the Great Depression of the 1930s and prompted by the challenge to liberalism led by an economist named John Maynard Keynes (Martinez and García 2000) neoliberalism emerged and attempted to chart a midway between the conflicting philosophies of classical liberalism and collectivist central planning. In the 1960s, usage of the term 'neoliberal' heavily declined. When the term was reintroduced in the 1980s in connection with Pinochet's regime the usage of the term neoliberalism had shifted. It had not only become a term with negative connotations employed principally by critics of market reform, but it also had shifted in meaning from a moderate form of liberalism to a more radical and economically libertarian set of ideas, reminiscent of the economic model developed by the 'Chicago school' economists in the 1960s and 1970s.

Neoliberalism in principle applies market metaphor in the perception of the world. But in practice it is more of an attitude than an economic reality. Most of its attributes are presented in definition of terms. By 1989, the ideology of neoliberalism was enshrined as the economic orthodoxy of the world. In other words, it assumed a hegemonic outlook in spite of the issues it has with crony capitalism.

Neoliberalism is certainly a form of free-market neoclassical economic theory, but it is quite difficult to pin it down further than that (Mohammadzadeh 2011). In a conscious attempt to dissociate from critical literature on neoliberalism Thorsen and Lie (n.d.) in their article 'what is Neo-liberalism' explored the definition of neoliberalism. They did not go beyond expounding free enterprise with minimal state intervention. They argued that free enterprise in their view can be obtained under the auspices of autocrats as well as within liberal democracies. Meanwhile there are indications in literature that neoliberalism is a form of neocolonization that has nothing to do with liberalism.

The neoliberal urban vision was adopted, without debate, by many city governments in the 1990s. Urban planning in the neoliberal era has to contend with rollback government intervention and rollout market mechanisms and competition. Neoliberalism dictates these preconditions in an attempt to combine classical liberalism and the theory of growth contained in Keynesianism-controlled liberalization. A matrix of neoliberal policy settings is presented in Table 4.1 to represent the structure of neoliberal theory that informs paradigm shifts in urban planning in contemporary political economies. This explains the ideological foundation for

4.3 Neoliberalism as Meta-Theory of Planning

Table 4.1 Matrix of neoliberal theory from planning perspective with political implications

Matrix			Criteria	Political implications
	Objectives	Options		
Policy framework	Government function	Decentralization	Reduced central government planning	Greater electoral accountability
				Stronger role for local clientele relationships
		Depoliticalization	Limited central government control of local government planning	
		Agencification	Contracting out of planning functions	Fiscally constrained local governments
			Central government off-loads unfunded risks and responsibilities to local governments	
			Policy solutions borrowed or adapted across jurisdictional boundaries	
	Government policy focus	Liberalization	Focus on innovation and competitiveness rather than on full employment and planning	Reduced social cohesion
			Social wage is seen as a cost of production rather than as a means of redistribution to maintain social cohesion	Increased social exclusion
			Welfare to work to reduce welfare Expenditure	
	Economic management	Financialization	Less maintenance of infrastructure and services	Fiscally constrained governments
		Fiscal conservatism	Limited provision of infrastructure and services	Infrastructure and services failures
			Greater private sector provision	Price hikes
			Reduced developer contributions in new growth areas	Cross-subsidies are increased
			Reduced focus on urban renewal projects	Rent seeking by the private sector
			Focus on cost recovery and user Pays	
	Government regulation	Deregulation	Removal of comprehensive master planning and collaborative planning policies and practices	Less importance of rules, processes and expert jurisdictions
			Simplified planning regulation	Less concern for development externalities
			Plans that are more flexible	Stronger role for the private sector in determining the form and location of development
			Plans that give less direction to local government	
			Plans that give more certainty and predictability to developers	Potential impact on the spatial cohesion of cities
			Plans with fewer directives and more negative regulation	
			Plans that specifically integrate central and local government priorities	Reduced oversight and increased risk of corruption
			Enabling regulations for major or mega projects	Risk of regulatory capture
			Use of reserved planning powers (Ministerial call ins and directions) to facilitate projects	
			Speeding up of development assessment, public inquiry and plan preparation processes	

(continued)

Table 4.1 (continued)

Matrix		Criteria	Political implications
Objectives	Options		
Central and local government relationship	Growthism Entrepreneurism	Local governments focus on place branding, marketing, promotion and competition rather than place making	Local governments forced to compete with each other for economic growth
		Local governments focus on economic growth projects generally in central city locations at the expense of investment elsewhere	Reduction in public services
		Politicians and planners gain financial acumen and act as urban entrepreneurs	
		Governments mimic corporate style and logic	
		Public services seen as ineffective and wasteful and a drain on entrepreneurial activity	
Government and private sector relationships	Marketization and privatization	Rise of the intermediate services sector (professional advisers)	Loss of citizen entitlements
		Developer-led development rather than plan-led development	Excess profits
		Developers takeover plan making	Price hikes
		Developers are stakeholders in major public infrastructure projects	Asset stripping
		Public assets privatized or divested	The poor driven to the worst located areas
		Compulsory purchase of private land for public benefit by private landholders	Profit seeking by private contractors increases public sector expenses
		Business improvement districts (US) where revenue from a district is spent in a district	
		Privatized planning regulation (for example private certification)	
		Limited public review of public infrastructure projects (focus is on selling the project not evaluating the project)	
		Private sector involvement in financing and operating infrastructure	
		Competitive bidding for urban renewal and infrastructure projects	
		Private sector provision of rental housing rather than public housing	
		Privatization of public spaces (public plazas; pavements; urban parks; government land and buildings)	
		Privately governed and secured neighbourhoods through management (for example gated communities, community interest developments and Homeowners Associations in the US) and passive design (for example master planned residential estates)	
Government and civil society relationship	Individualism and clientelism/consumerism	Corporate style advisory boards replace community-based consultative groups Focus on private hospitals and private health insurance rather than public hospitals	Downsizing of services
		Focus on owner occupied and rental housing rather than public housing, community houses and housing associations	Limited access to shelter and services for the poorest
		Focus on private schools rather than public schools. TAFE and other public educational facilities	Rise in the informality in cities
		Limited investment in social infrastructure to address areas of social exclusion	

Source Own construction 2013 (Adapted from: Wright and Cleary (No Year) Are we all neoliberals now? Urban planning in a neoliberal era. Available at: www.herbertgeer.co.au Date of access: 2 July 2013 framework planning which discourages detailed provisions, leaving such details to neoliberals to determine in line with their philosophy of market metaphor.

The concept of neoliberal planning is built into the framework of neoliberal theory set out in Table 4.1. However, the phrase 'neo-liberal planning' is a paradox built on ideological contradictions. This is the inevitable conclusion of critical literature which holds that planning is abhorrent in neo-liberal ideology and combining them in one phrase 'neoliberal-planning' is discomfited, implying that planning 'beyond the profit principle has reached its limits in the 21st century' (Baeten 2012: 206). Mohammadzadeh (2011) summarized neoliberal planning as market-oriented planning practice, planning without plan. Mohammadzadeh further indicates that 'this model has largely reduced and misread planning concepts and techniques by merely creating greater financial benefit, regardless of the side effects of its practice in respect of social environmental impacts'. Neoliberal planning facilitates market forces in city management. In other words, it is reduced to communication in line with building a neoliberal society in which public interest is an illusion. Indeed neoliberalism attacks planning practice directly. It constrains what planners do and what they think they can do. Planners confront neoliberalism not only in practice but also in theory. Many courses in planning schools today revolve around the assumptions and abstractions of neoclassical economics (Goonewardena 2007).

Baeten (2012) provided a compendium of critical views on the concept of neoliberal planning. Among them neoliberal planning is portrayed as seeking growth that is individualistic and not collective hence it disconnects with the traditional growth principles of form and function in planning and uses Darwinian survival logic to resort to economic reductionism (Baeten 2012: 209), to suggest planning subjectivity and this draws from the mindset of freedom and not equity, to provide market-space economy with considerable impetus over land use management, and so on. The resultant neoliberal urbanism compelled geographers and planners to generate alternative concepts such as the 'just city' concept, the 'right to city' logic and perhaps the 'good city' concept, targeted to break the neo-liberal hegemonic outlook of neo-liberal ideology in city development—see Leys (cited in Peck et al. 2009).

4.4 Conclusion

In summary, neoliberalism as meta-theory for planning is contestable. The neoliberal planning it supports tends to have conceptual problems given its disposition to market-driven project development and bankability, and private profitability. It is more of investment planning than urban planning instrument. It alienates the state from undertaking non-profit-oriented project development. This perhaps explains why the entry point for neoliberal transformations is crisis locations, for instance at the inception of majority rule in South Africa, in Montreal during the de-industrialization period, in Russia at the fall of communism, in

Argentina when their economy collapsed into chaos between 1999–2002, in Nigeria during economic recession in the 1980s, etc.

There is no indication of relationship between the precepts of traditional development ideologies and the international perspective particularly neoliberalism. Thus a widening gap exists between the mission of planning driven by neo-liberal meta-theory and the traditional philosophy of development. A lacuna exists in the meta-theory of planning for the delivery of African renaissance.

References

Anarcho (2005) Neoliberalism as the new mercantilism. Available at http://www.anarkismo.net/article/192 Date of access 17 Feb 2015

Baeten G (2012) Neoliberal planning: does it really exist? In: Tasan-Kok T, Baeten G (eds) Contradictions of neoliberal planning. Department of Social and Economic Geography, Lund University, 22362 Lund, Sweden pp 206–210

Coleman DC (1969) Revisions in mercantilism. Methuen, London

Cornelli EM (2012) A critical analysis of Nyerere's Ujamaa: an investigation of its foundations and values. (Being a doctoral thesis submitted to The University of Birmingham. Centre for the Study of Global Ethics Department of Philosophy The University of Birmingham)

Davies R, Thurlow J (2009) Formal-informal economy linkages and unemployment in South Africa. International Food Policy Research Institute (IFPRI): Discussion Paper 00943

El Mahdi A (2002) Towards decent work in the informal sector: the case of Egypt. International Labour Office (ILO), Geneva

Gerxhani K (2004) The informal sector in developed and developing countries: a literature survey. Public Choice 120(3/4):264–294

Goonewardena K (2007) Planning and Neo-liberalism: the challenge for radical planning. Progressive Plann Magazine, 22 July

Hettne B (2014) Neo-mercantilism: what's in a word. Available at: http://rossy.ruc.dk/ojs/index.php/ocpa/article/viewFile/3924/2090, pp 205–229, Date of access 4 Mar 2015

Hicks J (1998) Enhancing the productivity of urban Africa. In: Proceedings of an International conference research community for the habitat agenda. linking research and policy for the sustainability of human settlement held in Geneva, 6–8 July

Mark T, Potter C (2004) Neo-liberalism, neo-mercantilism and multi-functionality: contested political discourses in a European post-Fordist rural transition. EPMG working paper. Available at: www.irsa-world.org/prior/XI/papers/2-8.pdf Date of access 17 Feb 2015

Martinez E, Garcia A (2000) What is "neo-liberalism?"—a brief definition. Available at: http://www.globalexchange.org/resources/econ101/neoliberalismdefined Date of access 18 Aug 2013

Mohammadzadeh M (2011) Neo-liberal planning in practice: urban development with/without a plan (A post-structural investigation of Dubai's development). Paper Presented in Track 9 (spatial policies and land use planning) at the 3rd World Planning Schools Congress, Perth (WA), 4–8 July 2011

Nyerere JK (1962) 'Ujamaa' the basis of African socialism. Dar es Salaam, April

Peck J, Theodore N, Brenner N (2009) Post-neo-liberalism and its malcontents. Antipode 41(1):94–116

References

Raza W (2007) European union trade politics: pursuit of neo-mercantilism in different arenas? In: Becker J, Blaas W (eds) Switching arenas in International trade negotiations. Ashgate, Aldershot

Thorsen E, Lie A (No Year) What is neo-liberalism. Available at: http://folk.uio.no/daget/neoliberalism.pdf Date of access 17 Aug 2013

Chapter 5
Theoretical Frameworks

Abstract The world system is theorized as an imperial instrument that retains Africa in dependent capitalism. Its operationalization perhaps explains the stagnation in the development of urban (structure) theory and the dynamics in the development of regional development theory. The review of regional development theories from 1930s shows that regional theories sought for the economic bases of integration. Thus, the theories synchronized with the ethos of the world system vis-à-vis market economy. However, the need for spatial integration persists and is perhaps responsible for the cyclical evolution found in urban planning theory, from classic in the 1960s to rational in the 1970s to neoliberal in the 1980s–1990s and then to neoclassical since 2000. Simultaneously, tendencies of recourse to spatial models of regional integration are gaining momentum.

Keywords World system · Market economy · Core-periphery · Inequality · Dependency theory · Productivity

5.1 World-Systems Theory

The world-system theory has been closely associated with Immanuel Wallerstein. This theory is a macro-sociological perspective that seeks to explain the dynamics of the "capitalist world economy" as a "total social system" (Martínez-Vela 2001). Wallerstein used three major intellectual building blocks to expound the theory, namely the Annals School, Marx, and dependence theory. Because, it placed a lot of emphasis on development and unequal opportunities across nations the theory is in many ways an adaptation of dependency theory (Chirot and Hall 1982). However, it picked its historical approach from the *Annales* School, whose major representative is Fernand Braudel. The impact of Marx bears more on Wallerstein's ambition to revise Marxism itself. Since it is from a dependency theory perspective that many contemporary critiques of global capitalism come from, the theory is embraced by development theorists and practitioners.

What is a world system? Martínez-Vela (2001) indicates that a world system is what Wallerstein terms a "world economy", integrated through the market rather than a political centre, in which two or more regions are interdependent with respect to necessities like food, fuel, and protection, and two or more polities compete for domination without the emergence of one single centre forever (Goldfrank 2000). Initially in 1974 Wallerstein defined the world system as a 'multicultural territorial division of labour in which the production and exchange of basic goods and raw material is necessary for the everyday life of its inhabitants'. Two years later Wallerstein (cited in Martinez-Vela 2001: 3) explained the world system to mean:

> A social system, one that has boundaries, structures, member groups, rules of legitimation, and coherence. Its life is made up of the conflicting forces which hold it together by tension and tear it apart as each group seeks eternally to remould it to its advantage. It has the characteristics of an organism, in that it has a life-span over which its characteristics change in some respect and remain stable in others…Life within it is largely self-contained, and the dynamics of its development are largely internal.

The concepts of world system also borrowed from the Hungarian economic anthropologist Karl Polanyi who identified three modes of economic organizations: reciprocity, redistribution and the market (Wallerstein 2004). The reciprocity mode creates social bond and begets 'Mini-systems', the redistribution mode creates group solidarity through central control and begets 'World-empires' while the third category, self-regulating market exchange conducted by independent action, related by contract, profitability alone to determine what kinds of action survive creates individualistic "accumulation without end" and begets 'World-economies'. The three outcomes—Mini-systems, World empires, and World economies—are historical systems. The mini-system has within it a complete division of labour and a single cultural framework as found in traditional societies.

The world systems are units with a single division of labour and multiple cultural systems, the world-empires have a common political system—for example, the Roman Empire—whereas world-economies do not have a common system—for example, the world today (Wallerstein 2004). Given the aforementioned insight Wallerstein (2004) further defines the term world system thus:

> … it is a spatial/temporal zone which cut across many political and cultural units - including different political states and peoples - that represents an integrated zone of activity and institutions which obey certain systemic rules.

Note that gradually the definition of World system is edging closer to spatial coverage in a scenario where mini-systems and world-empires recede in significance in theory because in reality reciprocal relations underpinned by networking amongst core countries persists (see Van Hamme and Pion 2012). However, and remarkably, Wallerstein indicates that world economies have historically been unstable, leading either towards disintegration or conquest by one group and, hence, transformation into a world-empire. He further indicates that only a capitalist world-economy can survived for such a long time as it is actually doing.

In a World-systems research perspective, the political state is replaced by historical systems. World-systems research is largely qualitative; however, it is conducted in an integrated disciplinary context, which in social science literature is referred to as idiographic methodologies. The world-system approach explains a strong tendency towards spatial polarization at the world scale and its persistence over time (Van Hamme and Pion 2012: 67). By using methods in line with the worldsystem and dependence theories economic flows—trade and foreign direct investment—still deeply separate core and peripheries (Van Hamme and Pion 2012: 65).

Core and periphery are not only characterized by their level of development but also by unbalanced relationships, which in turn explain unequal development and its persistence over time; indeed, the core has imposed successive economic specializations—in raw materials—to the periphery for the needs of its own accumulation and this has introduced a complex dialectic between endogenous capacity of development and external relations (Van Hamme and Pion 2012: 3). On account of this development perhaps Wallerstein (cited in Van Hamme and Pion 2012: 67) introduced the concept of 'semi-periphery' to break the dualistic division. According to Steiber (1979: 24) 'the semi-periphery shares some of the benefits accruing to core status by their exploitation of the periphery, but they are still exploited, in turn, by the core'.

Evaluating core-periphery relations highlights the dependency theory. This theory is 'a neo-Marxist explanation of development processes, popular in the developing world among whose figures is Fernando Henrique Cardoso, a Brazilian (Martinez-Vela 2001: 3). The core-periphery concept explains the functioning of the dependency theory: inequality borne out of necessity in established symbiotic relations between economic regions somehow skewed in favour of the dominant region. According to Dependency theory as represented by Wallerstein (1976) the structure of Third-World societies and their cities was primarily a result of the manner in which they had been integrated into an international capitalist system from mid-fifteenth century, and the continuing pattern of integration which as mentioned earlier was responsible for a particular socio-economic formation—peripheral capitalism which is very different from the capitalist mode of production in the developed capitalist countries (see Chukuezi 2010). In other words, their economies are located at the periphery of the international capitalism. These economies are structured to meet the needs of the capitalist system.

5.2 Growth of Urban (Structure) Theory

Classic theories of urban land use structure endemically represented by the concentric zone theory, sector theory and multiple nuclei theory have undergone repeated reviews in urban planning studies because of the poor dynamics in the development of urban theories. It is recalled that these sets of theories drew from Von Thunen's regional land use model for the analysis of agricultural land use patterns in Germany which developed in 1826. Remarkably, the issues raised by these out-dated models remain resilient in contemporary urban dynamics. Von Thunen's model introduced the distribution of land use around a spatial system following the universal law of growth which acknowledges nucleation in isolated state although his work assumed a unified agricultural space. Commonplace interpretations of the model ascribe the land use pattern to location economics that presume the community form to represent a market. This attracted transportation costs arguments especially for the modified version of the model which again presumed linear development along perhaps navigable river and the presence of competing centre.

In the early twentieth century Burgess (1925) introduced the concentric zones model with intent to analyze social classes. He adopted the perception of urban environment that is limited to the core area in which he applied Von Thunen's concentric concept. The dynamics of the model assumes a relationship between the socio-economic status (mainly income) of households and the distance from the Central Business District (CBD). This applies under formal conditions but under informal conditions where the processes of gentrification and succession are irregular as determined by smart growth processes the scenario changes. The fission of family compounds, if related to stages of development, could follow the concentric zone argument, but suburbanization characterized by low-cost housing schemes and the development of peripheral slums for urban migrants challenge the urban structure proposed by the model.

Subsequent theories discontinued consideration of the concentric zone pattern and introduced sectors. Hoyt (1939) maintained that the land use pattern was not a random distribution, nor sharply defined rectangular areas or concentric circles, but rather sectors. His sector model was conceived explicitly under the influence of a transport axis following the study of residential areas in America. More than five decades later Jean-Paul Rodrigue of Hostra University, USA upheld the notion that the model added the effect of direction and time to the effect of distance. He argued that transport has directional effect on land uses because transport corridors, such as rail lines, public and major roads, are mainly responsible for the creation of sectors.

Otherwise sector development derives from agglomeration that is also heavily influenced by economies of scale and the direction of innovations or even socio-cultural imperatives such as land ownership system. The land use clustering resulting from these influences do not necessarily take their bearing from the CBD as the model suggests and this perhaps informed the multiple nuclei model Harris and Ullman (1945) developed on land use and growth. Both models are illustrated in Fig. 5.1.

Harris and Ullman's (1945) multiple nuclei model introduced the integration of a number of separate nuclei in the urban spatial structure. It dropped the idea that a city develops from a one central business district. Rather cities develop from several points, each point acting as a growth centre for a particular kind of land use and later they merge to form a single urban area. Some of these nuclei are pre-existing settlements normally found in the outskirts of the city, sometimes near valuable housing areas, although the nuclei are not located in relation to any distance attribute. The distribution of sectors according to Jean-Paul Rodrigue depends on differential accessibility, land use compatibility, land use incompatibility, and location suitability.

5.2 Growth of Urban (Structure) Theory

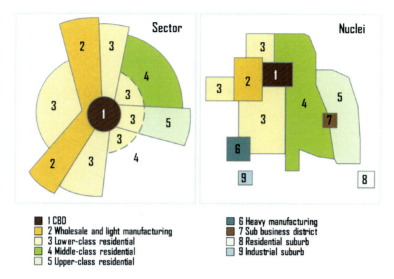

Fig. 5.1 Schematics of Homer Hoyt (1939) sector theory and Harris and Ullman (1945) multiple nuclei theory. *Source* Adapted from H. Carter (1995) The Study of Urban Geography, (Fourth Edition), London: Edward Arnold, p. 126

Within a decade of the transition from concentric to sector theories of land use structure attention moved from land use distribution to land use patterns, all within the ambit of sectoral development. Walter Isard's (1955) hybrid model (see Fig. 5.2) introduced this transition in which the concentric effect of nodes and the radial effect of transport axis mix to form a land use pattern. Land rent theory extended from here and moved more into the frontiers of land economics than land use structure. Ever since nothing much has happened concerning urban land use structure, theories, except for some strategic theories that are beginning to sprout in the transportation planning sector which attempt to explain the urban spatial system with transportation as the primary dynamic element. Okosun (2013) cited some scholars (including Anas 1980, Landis 1995, Batty 2002, and Pettit 2002) who have all developed modern theories for understanding urban land use patterns.

The processes of land use change and the resultant configuration of urban form is a theoretical phenomenon that requires much more than market consideration of transport costs to comprehend, although transportation perspectives could be the bridge that will link regional development and urban land use planning theories. The extent to which transportation perspectives could play this role lies in how it compares with urban planning perspectives on matters arising in the development of spatial theories. My own assessment of both perspectives is outlined in Table 5.1. Transportation planning perspectives derived mostly as presented in the 1998–2007 periods by Dr. Jean-Paul Rodrigue, Dept. of Economics & Geography, Hofstra University, New York, USA, while those of urban planning perspective are educated guesses.

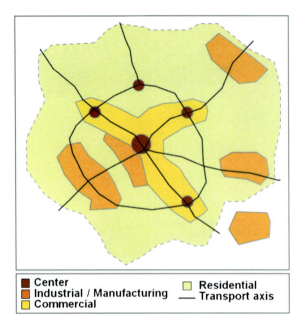

Fig. 5.2 Schematics of Walter Isard (1955) Hybrid land use model. *Source* Copyright © 1998–2007, Dr. Jean-Paul Rodrigue, Department of Economics & Geography, Hofstra University, New York, USA

Table 5.1 Matters arising from the development of spatial theories

Themes	Transportation perspective	Urban planning perspective
Economic growth	Consider movement costs	Consider infrastructure costs
Activity type categories	Routine, Institutional, and Production	Service, Manufacturing, and Primary (formal & informal)
Economic activity base	Formal employment	Livelihood support system
Urban environment	Uniform spatial surface	Manifold spatial systems
Urban change	Determined by movement patterns	Determined by growth dynamics & physical change
Urban form	Geographic patterns	Land use (or activity) system, distribution and patterns
Dynamics of urban form	Land economics	System of base: environmental factors, cultural base, value system
Land use	Determined by transport demand	Determined by function
Land use relationship	Accessibility: flows of passengers and freight	Functional flow of activities
Equity	Economic cost	Spatial integration
Efficiency	Maximize connectivity to minimize transport cost	Form and function

Source Own construction (2013)

The themes considered have conceptual meaning, which the transportation perspective did not fully capture compared with the urban planning perspective. The positions of transportation planning are rather skewed and subjective and may not lead to a general theory. Nevertheless transportation systems and spatial interaction have close relationships that impact on land use systems, distribution and patterns. The changing combination of these interacting elements underpins urban dynamics. Predicting and containing these dynamics in the urban system as regional development and urban planning theories provide impetus and this is the subject matter of planning paradigms. Traditional land use planning is known to apply form-and-function principles but that strategy is under contention much in the same way that efficiency intervention is in the equity versus growth argument. Thus a paradigm shift is underway in a neoliberal dispensation.

5.3 Growth of Regional Development Theory

This sub-section provides insight into the concept of region and regional integration. Subsequently it reviews contributions of new knowledge in regional development theories. The choice of theories to be investigated stands on the existing protocol established by Dawkins (2003) in his overview of the literature on regional economic growth. Dawkins (2003: 132) in his words 'reviewed seminal works and comprehensive overviews of the most important theoretical concepts'. This accepts his approach and adopts the same set of regional theories for investigation—however, from a different intellectual perspective. Therefore his work is the primary referral document used in this review. The links and relationships between these theories will be examined with the intent to identify insights that have the potential to inform further contributions towards a general theory of integrated regional development.

5.3.1 *The Concept of the Region and Regional (Economic) Integration*

The objective concept of regions crystallized following the development of trade and growth theories in the 1930s. Within the objective concept three categories of regions are recognized, namely, nodal (formal) regions, functional regions, and planning regions. The formal region is further classified into 'physical formal region' and 'economic formal region'. The formal region is the arena for new urbanism, especially a physical formal region that is linked with the concept of geographical determinism. This concept initiates lots of constraints and raises lots of concern for spatial planning. The economic formal region that represents types of industries or agriculture and the functional region sometimes referred to as nodal or polarized region are critical units for comprehending the 'city-region' in spatial planning. The functional region is understood to be a geographical area composed of heterogeneous and interdependent units. These units are the economic formal regions, cities, towns and villages. They display a functional coherence revealed in the form of flow of socio-economic activities. The functional region is preferably delineated through flow analysis. Typical examples include the polycentric region and extended metropolitan region.

The planning region comprises formal or functional regions or a combination of both. The planning region covers a wide geographical area and is defined as a "region that must be large enough to take investment decisions of an economic size, must be able to supply its own industry with the necessary labour, should have a homogeneous economic structure, contain at least one growth point and have a common approach to and awareness of its problems" (Glasson 1984).

Several other perceptions of regions exist. Some of these perceptions according to Dawkins (2003: 134) are in terms of "degree of internal homogeneity with respect to some factor (Richardson 1979), the size of a labour market, income, "sectoral specialization of labour (e.g., manufacturing-based regions versus service sector regions)" leading to the concept of homogeneous nations, "natural resource, ecosystem, or other geographic boundaries", and historically determined interdependencies between natural resource systems and human populations.

On the other hand regional integration indicates two perspectives: economic and spatial integration. Mostly what is discussed extensively in the literature is regional economic integration. Literature on regional spatial integration is a more recent phenomenon that emerged in the 1980s with the incidence of territorial planning subsumed in ESDP initiatives in Europe which are yet to fully crystallize. Regional economic integration is largely focused on trade agreements, collaborations and determined by the willingness and commitment of independent sovereign states to share their sovereignty. More or less it has become a political economy initiative invariably built on trade liberalization and organized either on a supranational or an intergovernmental decision-making institutional order, or a combination of both (see Van Langenhove and Costea 2007).

Overall, international trade is the basis for regional economic integration and countries are units of operation. Trade integration is the subject matter which is sought in neoclassic terms that presumes the convergence hypothesis drawn from either the Heckscher-Ohlin-Samuelson (HOS) theorem or the neoclassical growth model. Participating countries commit to measures of mutual co-existence, a provision that carries a dynamic convergence hypothesis although it is not clear how it works in the context of neoclassical trade theories. In the African context divergent economic indicators indict the convergence hypothesis. The core-periphery paradigm of world systems theory plays out instead, although by admitting economic growth as a path-dependent evolution the African experience can be explained with structuralist theories. Most African countries have been known to have engaged in structural adjustment programmes since the 1980s.

5.3.2 Review of Regional Development Theories (from 1930s)

Regional development theories underwent four phases of evolution within the period under review. The phases include: classic location theories (1930s–1960s), structural theories (1960s–1970s), institutional theories (1970s–1990s), and neoclassical theories (since the 1990s). Each phase comprises a set of theories triggered by an original contribution. Several authors contribute cumulatively in diverse directions however within the theoretical perspective(s) that identify each phase.

Classical location theories were the forerunners of regional development theories. The location theories were developed in the 1930s principally by Weber (1929) and Christaller (1933) who originated location theory and central place theory, respectively. Other contributions polarized into two perspectives. Those that focused on regional economic convergence hypothesis include the export base theory and exogenous growth theory, and those that focus on regional economic divergence hypothesis include cumulative causation theory and growth pole theory.

The development of growth pole theories seems to inhibit the influence of development ideology. Neoliberalism was taken as a constant and involuntarily dominated the mindset of theorists who were mainly economists and economic geographers. So it was difficult to pull out of the economic mainstream to consider integrated growth in the spatial systems. Spatiality was seldom conceived in terms of 3-dimensional space although towards the later part of the development of growth theories prospects grew with the contribution of Boudeville in (1966). Integrated growth is not only manifest in convergence of productivity as it is sought in spatial organization theories it is also manifest in regional spread. Regional spread requires interventions besides trade policies. Therefore, contribution in the context of growth theories should explore Boudeville's idea of considering geographic space and a viable alternative could be found in articulating growth theory based on activity distribution in formal and informal space and in different regional classifications.

Structuralist theories continued in the fashion of classical location theories to look at economics of growth. In this instance economic growth is viewed as a path-dependent evolution that passes through various stages of economic maturity. The evolution is normally triggered by socio-economic and political transformations, thus almost invariably suggesting structural adjustments in the economic system. This arena is a realistic experience for most spatial systems and it relates very well with the current wave of in-formalization experience that is sweeping through national economies in Africa. These sets of theories were pioneered by Hoover and Fisher (cited in Dawkins 2003: 140) but expanded mostly in the 1960s.

Unlike classic location theories structuralist theories are not modelled as intervention instruments. The progression from different perspectives it introduced is insightful and a good contribution. However, most of the analyses remain within the economics of profitability although with structuralist theories there is a little shift towards the realm of politics. A further shift is required to achieve a balanced synthesis of the analytical instrument. The history of economic geography is not disconnected with some measure of spatial planning intervention. Thus far the element of planning rationality in regional growth is not visible.

Institutionalist theories are coined to represent theories that deal with political institutions and regional economic development. These theories adopt interventionist perspective earlier identified to be the shortfall of existing theories. Hence, they are concerned with politicians and planners and their input in managing regional growth in their constituencies. This insight informed the Growth machine theory associated with Molotch (1976). In this theory reverse causality is found between regional growth and local political organization—regional growth is the platform for political organization and not vice versa. Because of the coalition of political organization of land-based elites which is involved, the growth machine theory is sometimes regarded as a theory of local politics. Logan et al. (cited in Dawkins 2003: 146) related growth machine to economic outcomes in a review of literature in growth machine theory. The outcome of the review somehow indicates that growth machines do not seem to impact the distribution of economic activities because neither pro-growth policies nor growth-control policies are effective. These are issues for growth machine theory to resolve for it to retain its status as a theory of how growth coalitions affect regional economic outcomes otherwise it is more useful as a theory of why political coalitions form.

Neoclassical theories are the latest contributions to the development of theories. They focus on addressing criticisms levelled against classical exogenous growth theories. This set of theories is endogenous growth theories and unlike classical exogenous growth theories hold that the source of growth of the region is not external to the region. The latest contributions in endogenous theories looked at the impact of infrastructure investment on regional productivity.

The other category of neoclassical theories is the new economic geography associated with Paul Krugman. Krugman's primary contribution in this model is to incorporate external scale economies and increasing returns into traditional models of interregional trade. He further introduced the core-periphery model to explain the emergence of clusters of economic activities due to a combination of centrifugal and centripetal forces. Krugman (1999) identified two approaches of the new model. The first 'considers the role of geographic factors such as climate and topography in determining patterns of regional growth and decline' and the second enquires into explanations for different 'patterns of economic growth when there are no apparent geographic differences between regions. Krugman (1999) combined both approaches to determine why differences in natural geographic features across regions can have such large persistent effects over time'. This set of theories provided for the systems of cities, different industrial structures, and different patterns of land use to emerge endogenously.

Details of contributions at the various phases are summarized in Table 5.2.

The main arguments of traditions in location theories are summarized in Table 5.3. It is observed that the element of transportation cost continues to maintain systemic relevance as regional development theories continues to evolve. Since early 2000 the tendency of integrating existing theories is underway. Notable

5.3 Growth of Regional Development Theory

Table 5.2 Review of regional development theories

Category of theories	Theorists	Contribution(s)
Classical location theories (1930s–1960s)	Alfred Weber (1929)—Location theory	**Location theory** • Introduced transportation cost and labour wages as primary location determinants for the distribution of industrial function in urban space • Used agglomeration and de-glomeration principles—local factors • Engaged functional flow of market operations within the spatial system • Critics reasoned that the economy of scale and pricing behaviour of firms are known to influence the location of industries besides other factors of production such a land viewed in economics as geographic location and capital asset • Walter Isard (1956) drew from location theory concept to introduce regional science
	Walter Christaller (1933)—Centre place theory	**Spatial organization theory** • Bifurcates urban space into central places and catchment areas • Introduced the level of specialization of towns as criteria to produce "tiered" spatial systems in the distribution of central places in urban space • Description of spatial behaviour that gives indication of centripetal forces in urban space • Spatial segregation of manufacturing and service functions • Losch (1954) puts the initial idea of Christaller in an economic context and postulates transportation, cost of operation, and market as determinants of spatial organizations • Christaller (1966) introduced marketing, transportation and administrative principles in establishing central place patterns
	Charles Tiebout (1956a, b) and Douglass North (1956, 1955)—Export base theory	**Growth theory (convergent hypothesis model)** • Introduced demand-side approach to regional growth, that is region's response to exogenous world demand

(continued)

Table 5.2 (continued)

Category of theories	Theorists	Contribution(s)
		• Bifurcates industries into basic and non-basic industries • Introduced multiplier effect principles • Considering growth independent of manufacturing • Introduced supply-side factors incidental to population large enough to affect world demand for exports • Using supply-side models of investment in regional productive capacity. This is responsible for neoclassical exogenous growth theories developed by Solow (1956) and Swan (1956)
	Gunnar Myrdal (1957)—Cumulative causation theory	**Growth theory (Divergent hypothesis model)** • Introduced self-regeneration principles in regional growth • Introduced 'developed' and 'lagging regions' concepts • Introduced 'spread-effect' and 'backwash-effect' concepts • Identified functional relationships between regions—forward and backward relationships • Kaldor (1970) contributed the consideration of 'efficiency wage' concept • Dixon and Thirlwall (1975) contributed the consideration of 'Verdoorn effect' concept
	Francois Perroux (1950)—Growth pole theory.	**Growth theory (Divergent hypothesis model)** • Introduced the perception of space as force • Introduced the functional interaction between firms and industries • Hirschman (1958) contributed the 'trickling-down effect' concept • Boudeville (1966) contributed contextualizing in geographic space • Friedman (1966) contributed the core-periphery model • Friedmann (1972) introduced authority-dependency relations in spatial systems

(continued)

5.3 Growth of Regional Development Theory

Table 5.2 (continued)

Category of theories	Theorists	Contribution(s)
		• Richardson (1973) introduced space and distance
Structuralist theories (1960s–1970s)	Stage/sector theory	• Perloff et al. (1960) identified three-stages of change of growth of the regions (in America) • Hoover and Fisher (1949) identified four stages of growth that indicated internal changes in the division of labour that produce economic specialization • Rostow (1977) identified five-stages of growth and introduced the "Take-off" period concept (originally suggested in 1960) • Pred (1977) introduced "Spatial biases" concept • Thompson (1968) introduced "Ratchet effect" concept
	Profit/product cycle theory	• Markusen et al. (1999) introduced the role of large firms, state actors, local fixed capital, and the active recruitment of skilled labour in district formation • Scott (1992) and Cooke and Morgan (1993) looked at the control advantages of hierarchical forms of transaction governance • Porter (1990) looked at geographic clustering • Markusen (1985) contributed profitability and identified five 'profit cycles' with his profit cycle theory • Vernon (1966) introduced the product cycle concept
	Marxist theory	• Argued the spatial dimension of division of labour (Massey 1984) • Introduced "Regional rotation" concept (Goodman 1979) Argued geographic displacement of labour (Holland 1976)
Institutionalist theories (1970s–1990s)	Growth machine theory	• Adopts interventionist perspective • Argues that regional growth is the platform for political organization
	New institutional economics	• Introduced the concept of transaction cost of production • Introduced institutional adaptation and change (North 1990)

(continued)

Table 5.2 (continued)

Category of theories	Theorists	Contribution(s)
Neoclassical theories (since 1990s)	Endogenous growth theories	• Argues that the growth of the region is internal • Incorporates a savings rate that is determined by household choice • Schumpeter (1947) introduced the process of innovation dimension • Arrow (1962) introduced the 'learning-by-doing' argument • Romer (1986) incorporated technical change as an endogenous parameter • Barro's (1990) model incorporates tax-financed public services • Theorists considered the impact of infrastructure on private sector coordination
	New economic geography	• Incorporation of external scale economies • Considers the role of geographic factors • Enquires into explanations for different patterns of economic growth when there are no apparent geographic differences between regions • Krugman (1999) introduces the spatial dimensions of regional growth and trade

Source Own construction (2013)

efforts are readily available (see Acs and Varga 2002), Bretschger (1999), Fujita and Mori (1998), etc.). These integration efforts sought an explanation for regional economic growth. Also underway are efforts to extend endogenous regional growth theory vis-à-vis the role of leadership and pioneered by Friedman (1966). Besides leadership, the element of efficiency intervention in the name of equity is alive in public debate vis-à-vis its justification in the light of market failures.

Efficiency and equity tend to conflict or trade-off on each other with the result that compatibility between efficiency and equity is highly contentious. Although there are avenues of compatibility explained by theorists that highlight the stages of development, the utilization of regional resources, and the objectives of development especially in core-periphery regions where compatibility can be found between efficiency and equity goals are visible. Ultimately equity need not to be considered separately from efficiency as seems to be the case in endogenous growth theories.

Overall a lot of work has gone into the development of regional development theories. Regional growth in economic terms is at the centre of most contributions.

5.3 Growth of Regional Development Theory

Table 5.3 Main arguments of traditions in location theories

Theory	Main argument
Neo-classical	Location subject to free market forces
Behavioural	Behaviour of individual business. Decisions are made with limited information. Sub-optimal location choice
Institutional	External factors such as values and institutions. Mergers and acquisitions
Economic base	Related to the export industries of a region
Location factors	Specific location factors. Agglomerations of economic activity. Regional characteristics
Cumulative causation	Upward spiral where success breeds success (lack of success can lead to a downward spiral)
Core-periphery	Regional functions. Relationships between core regions and peripheral ones
Industrial district	Focus on networks, entrepreneurship, innovation, co-operation, flexible production and specialization
Innovative milieu	Importance of the cultural and institutions (synergies among local actors which give rise to fast innovation processes)
Competitive advantage	Competition between locations subject to factors related to labour, energy, resources, capital as well as proximity to markets

Source Adapted from McQuaid et al. (2004) The Importance of Transport in Business' Location Decisions, Department for Transport. http://www.dft.gov.uk/stellent/groups/dft_science/documents/pdf/dft_science_pdf_027294.pdf

It would have been interesting if theorists in their analysis that contributed to stage/sector theories considered the stages in the history of African civilization marked by major socio-economic and political transformation, or considered the form and function principles of growth in spatial planning in other to carry not just efficiency and equity but also spatial form which informs functional flow and by implication transportation costs. The appraisal of transportation costs in isolation of transport design is deficient. Needless to say that transport design in turn has implications for land use distribution and patterns without which the regional development theories tend to disconnect from urban development theories. This is considered a critical oversight since the spatial system which the theories attended to comprise the urban areas or perhaps the core areas. This is why the development of the models is found to be lopsided in favour of neoliberal principles of growth and on that ground is not considered a general theory from spatial planning perspective. It is therefore not clear how regional development theories connect with urban land use structure theories. It is most likely that disconnect exists between the two bodies of theories.

Mindful of the fact that the behavioural patterns of individuals, institutions and firms have an imprint on land use, subsequent reviews of urban planning theories will focus attention on the relationship between transportation, land use systems, distribution and patterns and then urban change. This body of knowledge needs to connect with theories of regional development for a better understanding of spatial integration.

5.4 Growth of Urban Planning Theory (from 1960s)

A cyclical revolution is identified—from classic in the 1960s to rational in the 1970s to neoliberal in the 1980s–1990s and back to a neoclassic planning perspective since 2000. Participatory planning paradigms assumed orchestrated popularity since the 1970s but product-planning remained resilient and tended to rebound since the 1990s. This trend is perhaps linked with tendencies towards spatial integration in space as sine-qua-non for economic growth indicated in institutionalist and neoclassical theories of regional development. Details of what transpired during each stage are summarized in Table 5.4.

Table 5.4 Summary of events in the evolution of planning perspectives (1960s–2012)

Period	Event	Major characteristic features
1960s	Classic planning perspective	• Emerged in the context of imperial capitalism • Problem situation: spatial distortion in urban entities, rapid urbanization, need for coordination arising from reorientation from environmental (geographical) determinism to interventionist approach, • Land use planning maximally applied at urban scale to shape the city • Statutory planning: development proposals that do not accord with planning controls, objectives and design standards can be refused under law • Planning remained steadfast on its regulatory status • Design-oriented activity strictly bound with securing adequate living space for the health and well-being of the people • Adopted master plan instrument, a brand of spatial planning *References* Ilesanmi (1998: 18), Eng (1992: 163); Owens (1994: 440); Kunzmann (2004: 384), Crawford et al. (2010: 91), Albers (1986: 18-26), Albrechts (2001: 1), Fischler (2000: 194); Pepler (1949: 103)
1970s	Rational planning perspective	• Early 1970s European critics such as Friedmann, Meyerson, etc. influenced the perceptions of planning • New planning concepts emerged, including advocacy, normative, incremental, and transactive planning • Planning was no longer limited to 'survey, analysis and plan' but extends to monitoring and reviews • Design status retained to shape the city • Modern master planning assumed integrated status that provides for and coordinates all of development activity • Renewed criticism against planning ensued in early 1980s in Africa on the heels of similar events in Europe in the late 1960s for paradigm changes to participatory planning

(continued)

5.4 Growth of Urban Planning Theory (from 1960s)

Table 5.4 (continued)

Period	Event	Major characteristic features
1980s–1990s	Neo-liberal planning perspective	• Renewed criticisms against the master plan instrument • Spatial planning concepts started shifting from product-oriented to process-oriented activities • Upgrade of public participation in planning • Deregulation of planning-decision making • Transition from statutory to non-statutory planning that applies on a wider scale • Spatial Planning emerged as an instrument for the management of change, a political process by which a balance is sought between all interests involved, public and private, to resolve conflicting demands on space • Spatial planning focused on the distribution of resource utilization • Integrated Urban Infrastructure Development Planning (IUIDP) attempted to provide an alternative form of planning, linking infrastructure development to planning • Master planning instrument remained resilient • *References* Todes et al. (2010: 415), Orange (2010: 55), Landman (2004: 163), Albers (1986: 22), Cilliers (2010: 73), Adams et al. (2006: 50), Kunzmann (2004: 385)
Since 2000	Neo-classical planning perspective	• New paradigms diffused from the global north • Gradual return of emphasis on the classical view of spatial planning • 'Communicative planning' a new planning paradigm that almost emerged was put on hold • Spotlights on new urbanism a post-modern concept in urban design. 'New urbanism' as it is called in America or 'Sustainable urbanism' advocates design-oriented approach to planned development • Resilience of master planning: even where the nature of 'forward planning' has changed, the basic principles of the underlying regulatory system, as well as a universal modernist 'image' of urban development, tend to remain" (Watson 2009) • New planning paradigms exert no influence on the regulatory system • Matters arising from new innovations in urban planning in Africa (see **Box 3.1** below) • Creative urbanism and new master planning emerges

Source Own construction (2012)

After more than three decades of experimentation with neoliberal planning perspectives, the problems in the urban milieu are being reinforced with new matters arising. A succinct list of the matters arising is given in Box 5.1 below. Renewed efforts towards addressing these matters arising tend to highlight indirectly the resilience of formal planning principles given the incidence of a new master planning approach which is making waves in China, etc.

Box 5.1 Matters arising from new innovations in urban planning in Africa

- Problematizing the right to the city (Merrifield 2011)
- Multiplicity of spatial planning approaches
- Problems with coordination—line function departments still function in isolation (Watson 2009: 180)
- Parallel planning systems (Halla 2002)
- Institutional deficiency—need for capacity-building (Harrison, Todes & Watson cited in Watson 2009: 180)
- Resilience of conventional (traditional) technique
- Professional participation (Watson 2009: 180)
- Participatory budgeting (Cabannes 2004)
- Disconnect with commoditization of land (i.e. land use management system)
- Temporarily unrealistic in the context of global south
- No legal foundation (except for few exceptions such as 'Integrated Development Planning' (IDP) that is backed up by national legislation in South Africa) (Parnell and Pieterse cited in Watson 2009: 179)
- New planning paradigms do not resonate with local institutions (UN-Habitat 2005: 5)
- Community engagement at the heart of planning practice (Brownhill and Carpenter 2007: 621)
- The hegemony of foreign planning systems (Okeke 2005)
- Older approaches of urban planning stand accused of being—anti-poor' (Watson 2009)
- Persistent urban problems (Werna 1995)
- Problematic assumption that liberal democracies can work in all parts of the South (Rakodi 2003)
- Complex and sophisticated system (Watson 2009: 180)
- No major impact on the process of urbanization in developing countries (UN-Habitat 2005: 5)
- New land management and tenure systems are required (Rakodi 2006)
- 'Brownfield' project based
- Planning cannot 'solve' the crisis of urbanization (Roy 2009)
- Land regulation systems remains constant (Watson 2009: 181)
- Conceptualization of public interest

Source Own construction (2011)

5.4.1 Master Planning Paradigm

In a neoliberal dispensation the words "master", "design", and "control" brand the master planning paradigm. They elicit immediate resistance aimed at reconsidering master planning outlook as a planning instrument. Neo-liberal mindset clouds the potentials of master planning and leaves zero tolerance to jettison it as an instrument for spatial organization. This position has the effect of relegating receptive views on master planning. Such tactics do not apply here on grounds of theoretical expedience. Planning and control are inseparable in the sense that unplanned action cannot be controlled in so far as control involves keeping activities on a predetermined course and rectifying deviations from plans.

Traditional master planning is synonymous with land use planning in space. In the 1960s, it subsumed socio-economic considerations and new concepts emerged, some of which are advocacy, normative, incremental, and trans-active planning. The new outlook of master planning expanded its objectives to include the renewal of the economic base of spatial systems. Its attribute as a futuristic instrument heightened as well as its concern for economic growth. In the early 1980s, its scope, following the contribution of Ratcliffe (1981), conceptually increased to include monitoring and review processes. Ultimately, master planning paradigm functions as an instrument for planning, design, land use regulation or management. Its mode of operation consists of land use budgeting (plan generation) and spatial planning—design or the art of facility distribution. What master planning illustrates is that the nature of planning requires comprehensiveness and a holistic view of problems (Ogbazi 1992: 145).

5.4.2 The Participatory Planning Paradigm

The participatory planning paradigm is specifically concerned with interactive participation in planning which implies the decentralization of planning decision. Presumed to be an instrument for sustainable development and a requirement for plan implementation participatory planning is conceptually different from public participation. While participatory planning is a process-oriented approach to participation where stakeholders are involved in decision-taking, public participation is blueprint or target-oriented where participation is passive and consultative, in which case the public does not get involved in decision-taking. Therefore, participatory planning is defined as an interactive set of processes through which diverse groups and interests engage together in reaching for a consensus on a plan and its implementation. It operates with the principle of mediation, unlike public participation that operates with the principle of consultation. Indeed there are wide-ranging typologies of participation in planning as shown in Table 5.5.

Participatory planning is an event-centred concept. Its use is foremost when mediation in conflict situations and legitimacy and civic identity in plan preparation are fundamental to plan implementation.

Participatory planning process is invariably affected by practical questions such as "Whose knowledge is brought into the process?", "Who are the actors who participate in the process?", and "What are the spaces where participation becomes important?" (IDS 2002: 14). Actors who participate in a planning process are not neutral. Power relationships become an issue for the divergent interests the actors are obligated to protect, some of whom are 'manufactured' participants, that is to

Table 5.5 Typologies of participation in planning

S/no.	Typology	Characteristics
1.	Manipulative participation	Participation is simply a pretence, with 'people's' representatives on official boards but who are unelected and have no power
2.	Passive participation	People participate by being told what has been decided or has already happened. It involves unilateral announcements by an administration or project management without anyone listening to people's responses. The information being shared belongs only to external professionals
3.	Participation by consultation	People participate by being consulted or by answering questions. External agents define problems and information gathering processes, and so control analysis. Such a consultative process does not concede any share in decision-making, and professionals are under no obligation to take on board people's views
4.	Participation for material incentives	People participate by contributing resources, for example labour, in return for food, cash, or other material incentives. [People] … are involved in neither experimentation nor the process of learning. It is very common to see this called participation, yet people have no stake in prolonging technologies or practices when the incentives end
5.	Functional participation	Participation is seen by external agencies as a means to achieve project goals, especially reduced costs. People may participate by forming groups to meet predetermined objectives related to the project. Such involvement may be interactive and involve shared decision-making, but tends to arise only after major decisions have already been made by external agents. At worst, local people may still only be co-opted to serve external goals.
6.	Interactive participation	People participate in joint analysis, development of action plans and formation or strengthening of local institutions. Participation is seen as a right, not just the means to achieve project goals. The process involves interdisciplinary methodologies that seek multiple perspectives and make use of systemic and structured learning processes. As groups take control over local decisions and determine how available resources are used, so they have a stake in maintaining structures and practices
7.	Self-mobilization	People participate by taking initiatives independently of external institutions to change systems. They develop contacts with external institutions for resources and technical advice they need, but retain control over how resources are used. Self-mobilization can spread if governments and NGOs provide an enabling framework of support. Such self-initiated mobilization may or may not challenge existing distributions of wealth and power

Source Bass et al. (cited in Geoghegan et al. 2004: 5)

say, proxies. The knowledge base of participants with regards to data appreciation and knowledge of planning processes bifurcates into 'formal' and 'informal knowledge' which both have their merits, thus signalling the politics of appropriate 'expert knowledge' that applies in participatory planning. The space refers to contextual issues that maybe physical or not such as formal or informally constituted bodies—councils or committees—that have different rules of engagement.

The participatory process presents a trade-off between efficiency and inclusiveness, depending, however, on the level of participation anticipated. Time pressure, the needs of the community, the skills and experience of those participating, and the nature of the intervention, among other factors, all help to dictate the actual shape of the process (Rabinowitz 2013). The deliberations in the process are presumably compelled to remain within the survivalist level especially in Africa due to obvious limitations factored in top-down mobilization of participants, in-formalization, manipulative tendencies as well as dearth of capacity issues that determine integrative and specialized planning such as green point precinct planning which sustainable development demands. Involvement in plan implementation, monitoring and review, incidentally, is not persuasive yet it forms the major plane of argument for participatory planning as a continuing process.

The participatory planning paradigm, unlike master planning paradigms, is not a general theory in the urban planning domain. To this end the attributes of participatory planning compared with conventional master planning as shown in Table 5.6 are insightful. The attributes of participatory planning confirm a new perspective. The attributes of participatory planning lack integrative theoretical capacity to interpret in space the central issues of economic growth and land use in regional development theories. It will be more useful to serve as platform for the review of master planning paradigm than to provide an alternative approach for the integration of spatial systems.

Conceptually, as the scale of planning increases, the participatory process reduces (see Fig. 5.3). In other words participatory processes in preparing local plans or site plans are likely to be higher than what they will be for preparing urban and regional plans. In the same vein the statutory nature of planning should increase as the scale of planning reduces.

There are three trends that are instructive in considering the merits of paradigm shifts in urban planning. First, there is the impression that planning activities are no longer limited to traditional land use planning which is misconstrued in some quarters to mean that land use determination is now outside the purview of planning; second, the impression that planning is restrictive therefore exclusionary, not realizing that controls provide the bases for freedom. In any case what civic freedom does in planning control is to restrict under normal circumstance; or could it be that it is the mode of planning practice that is restrictive as it was the case in pre-1994 South Africa; and third, it seems the argument in the planning arena has moved from the manner of participation to the scale of participation. Altogether these trends reinforce themselves to make a case against conventional approach to planning and to choose a neoliberal path for the way forward. Unfortunately the re-modelling of the space economy which is considered a major challenge of

Table 5.6 Attributes of traditional and participatory planning paradigm

Attributes	Traditional (conventional) planning	Participatory planning
Concept	Thematic—for application in urban and regional development	Generic—for universal application
Objective	Land use planning for economic growth	Involvement for conflict resolution and mediation
Technique	Design-oriented	Management-oriented
Expertise	Consortium of consultant professionals in urban and regional development mandated by government or community	Facilitator(s)/motivator(s)—either a private individual(s) or an organization(s)—with or without government or community mandate.
Requisite skills	Analytical and creative ability	Communication and negotiation ability
Motivation	Productivity of the urban region and the urbanity of cities	Legitimacy and civic identity of development projects
Mission	Development planning for urban and regional integration to enhance productivity and maintain strategic relevance in city development	Survivalist project development planning to harness local resources and satisfy community needs
Methodology	Comprehensive planning.	Disjointed incremental project planning
Process	Definite known steps for plan preparation and implementation	Indefinite number of steps depending on matters arising most of which are outside the purview of planning
Outlook	Control-oriented	Management-oriented
Perspective	Linear blueprint approach	Non-linear process-oriented approach
Initiative	Urban and regional planning	Local planning
Participation	Consultative in principle	Interactive in principle
Framework	Spatial master plan	Non-spatial guideline plan (action plan)
Status	Spatio-physical and statutory	Non-statutory (although Bolivia has a Law of Popular Participation)
Time-frame	Maximally nine months barring bureaucratic delays to prepare a master plan for a medium-size city	Not specific—mobilizing stakeholders and settling conflicts for a local community is time intensive and could take more than a year before actual engagement in generating the action plan
Activity system	Technical analysis—based on quantitative and qualitative data	Political analysis—based on subjective information (public opinion)
Critical mass element (i.e. core activity)	Land use planning	Public relations (pre-mediation and conflict resolution) in project planning

Source Own construction (2012)

5.4 Growth of Urban Planning Theory (from 1960s)

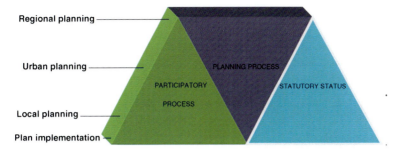

Fig. 5.3 Conceptualization of relationships in participatory cum planning process. *Source* Own construction (2012)

planning in Africa is a visionary activity in the realm of patriotism that lies beyond the domain of interactive instincts driven by private profitability as represented by the neoliberal planning perspective.

5.4.3 Theoretical Framework of Planning

Since the 1980s new perspectives in planning have introduced innovations in planning principles. As a result, a mixture of product-oriented and the new process-oriented planning principles tend to redefine the theoretical framework for planning. Based on an extensive review of literature, a summary of the emerging theoretical framework is presented in Table 5.7. The theoretical framework identified indicates that creative (form-based) planning principles are subjugated in favour of broad-based framework planning of neo-liberal dispensation.

5.5 Spatial Models for Regional Integration

There are spatial models of urban regional economic systems that have been theorized and found potentially useful in growing the economy of urban regions. Some of these models are developing informally through natural processes within spatial systems in Africa. They need to be identified and perhaps encouraged formally. Two of such models are reviewed below. These models are examined along with urban growth boundaries (UGB) as best practice mechanisms for growth management.

Table 5.7 Matrix of emerging theoretical framework for spatial and statutory planning in the 2010s

		Objectives	Options	Criteria
Theoretical framework	The concept of urban environment		Systems approach	**Study of system of base, deep structure, superficial structure**
			Spatial approach	**Delineation of urban core, urban outer ring (fringe area), urban outer ring (hinterland area)**
			Urban structure approach	**Concentric zone theory; sector theory; multi-sector theory**
	Statutory planning		Provisional documents	Framework plans
			Operational documents	Planning schemes; Layout schemes; Action plans
			Regulatory documents	Normative (informal) standards
	Spatial planning		Urban form	**Land use densities, patterns, functions**
			Urban growth management	**UGB, UDB, Greenbelt, Urban service limit, urban edge**
			Use of space (Land use control)	Market force as determinant factor
	Nature of planning		Informal	Non-professional planning; sectoraldevelopment planning; sub-area development planning
			Developmental	Project (facility) plans; economic plans
			Organizational	Stakeholders' forum; Event-oriented planning process
			Visionary	Long-term objectives; city vision statement; mind-set; outlook issues
	Purpose of planning		Economic	GDP, productivity; employment; use of resources; infrastructure planning, etc.
			Cultural	Conservation of heritage issues; cityscape concerns; civic identity concerns, etc.
			Health	Urban sanitation measures; urban quality control, etc.
			Form	**Urbanity standards, etc.**
	Planning instruments		Planning initiative	Urban planning and local planning approach, etc.
			Planning perspective	Process-oriented planning (neo-liberal planning)
			Planning framework	Guideline plans

(continued)

Table 5.7 (continued)

Objectives	Options	Criteria
Participatory process	Consultative	Professionals are under no obligation to take on board people's views
	Interactive	Groups take control over local decisions and determine how available resources are used
	Functional	Participation is seen by external agencies as a means to achieve project goals, especially reduced costs
	Manipulative	Participation simply pretence by representatives on official boards who have no power
	Passive	People participate by being told what has been decided or has already happened
	Self-mobilization	People participate by taking initiatives independently of external institutions to change systems
	Participation for Material incentives	People participate by contributing resources, for example labour, in return for material incentives
Planning methodology	Classic	**Survey-analysis-design approach**
	Rational	Participatory approach
	Neo-classic	New-master planning approach
Urbanism	New urbanism	Smart growth
	Sustainable urbanism	Design-oriented approach to planned development: relative density
	Creative urbanism	**Culture-based urban design**
Planning knowledge	Formal expertise	**Planning Professionalism; scientific database; use of planning theories & concepts, etc.**
	Informal expertise	Stakeholders' forum; town-hall meetings, opinion poll, etc
Plan evaluation technique	Planning as control of the future	Non-implementation of plan
	Process of decision making	Decision-making methodology; monitoring activities
	Intermediate technique	Plan alteration irrespective of implementation

(continued)

Table 5.7 (continued)

	Objectives	Options	Criteria
	Regional integration	Spatial integration	**Planning based on regional classifications: urban region, functional region, physical formal region, economic formal region, planning region, regional master plans, etc.**
		Space economy	**National urban development strategies (NUDS)**
		Regional connectivity	Regional road network; functional flow-chart
		Economic integration	Economic reforms; political reforms, etc.
	Cross-cutting issues	Heritage of city development in Africa	Civic identity
		Urbanization	Extroverted or introverted
		Urban growth	Spatial growth issues: qualitative, quantitative, structural growth, urban change, etc.
		Informal sector	Survivalist model
			Principles based on product-oriented planning

Source Own construction (2013)

5.5.1 The Extended Metropolitan Region (EMR) and Growth Triangle (GT) Model

The EMR is a development model that perceives transformative economic growth to occur congruently in a wider regional space than segregated into spatial constructs represented in urban or rural areas. It is a regionalization concept that adopts "region-based urbanization" instead of "city-based urbanization" to analyse the processes of economic and spatial development. So the EMR perspective is focused on deeper issues that extend beyond urban transitions and the urban-rural dichotomy that informs it to explore the wider arena of transitions in space economy. The EMR is therefore:

> Characterized by extremely high levels of economic diversity and interaction, a high percentage of non-farm employment (i.e. over 50 per cent), and a deep penetration of global market forces into the countryside. These regions may stretch for up to 100 km from an urban core and are frequently found near major transportation conduits (McGee cited in Macleod and McGee 1996).

In the EMRs, one finds that apparently rural areas are coming to adopt economic characteristics usually thought of as urban and industrialization and rapid development are coming to affect the people of these regions in situ and are also drawing in large numbers of migrants" (Macleod and McGee 1996).

Decentralization policy necessitates the emergence of EMRs. There are three types of EMRs, namely the "expanding city-state" as manifested in the Singapore EMR, poly-nucleated patterns of new towns associated with the situation in the Kuala Lumpur EMR, and a high-density EMR characteristic of Jakarta, Manila and Bangkok. In this third category, large rural populations polarize at the periphery of cities where their services are used for food crop production in the peri-urban areas and rural hinterland. This is the most ubiquitous EMR, perhaps modelled after the suburbanization of informal labour. However, the deconstruction of the urban concept, which is inherent in EMR model, raises serious questions about some central concepts in development theory such as the rural-urban social divide and traditional metropolis-hinterland models of spatial development.

The term "Growth Triangle" was coined in Singapore (in relation to Singapore-Riau-Johore GT) to convey the potential believed to be inherent in the synergistic interweaving of the comparative advantages of the three nations (Macleod and McGee 1996). The growth triangle (GT) model is an instrument for managing EMRs. It has the potential to trigger the concentration of certain activities in urban cores and the explosive transformation of the countryside although it is usually caught in the web of political manoeuvres especially the trans-national GTs. This informs the stress the GT girding Singapore is having over the appearance that sovereign areas of Malaysia and Indonesia are coming under the sway of Singapore (Macleod and McGee 1996).

5.5.2 The Poly-centric Model Compared with the EMR Model

The poly-centric model is based on the hypothesis that agglomeration economies of the primate city offer unequal advantages that necessitate the development of sub-centres or satellites linked by a metropolitan transport system that will often improve the efficiency of the metropolitan region, increasing its attractiveness as an industrial location and to migrants (Richardson 1981). But this model leads to large polarized metropolitan region that often conflicts with inter-regional equity goals (Richardson 1981). In the African context where political realities make equity a critical variable for the new economic and spatial entities sought in the African renaissance the application of poly-centric model without growth management mechanisms is in a disadvantage when compared with the EMR model. This position draws strongly on the positive outlook of an EMR towards extroverted patterns of urbanization which are identified as being compatible with equitable regional integration. It does not question the adequacy and appropriateness of the polycentric model for integrated development. Rather it simply draws attention to the leverage EMR has in equity matters.

Related issues here have to deal with scale of development and level of development. Large-scale development is expected to trickle-down from large to lower-scale nodes in the case of the poly-centric model. EMRs seem not to operate on this principle. It rather dwells on same level of development for all settlements regardless of rural or urban status. However, it all depends on the type of EMR model. Not all types of EMR are appropriate for leverage in attaining equity. The 'expanding city state' model such as in Singapore city-state and the 'high density' model that is characteristic of Jakarta, Manila, and Bangkok tend to have issues with equity if compared with the 'multiple nuclei' model found in Kuala Lumpur and Seoul regions where there is comparatively low population density with growth management systems. Kuala Lumpur has been able to control its growth through the creation of a poly-nucleated pattern of new towns and smaller suburban centres located along the major arterial routes of the city (Macleod and McGee 1996).

5.6 Conclusion

The study of theoretical frameworks drew attention to the world system which development in Africa is contending with. In line with neoliberal development ideology it seems the world is condemned to market economy. The impact of this context is phenomenal in the development of urban theories. The last known input in these sets of theories was indeed in mid-1950s with the contribution of Isard's (1955) hybrid theory followed three decades later in the 1990s by other emerging contributions. This provided the theoretical gap to leverage neoliberal planning theory. Hence, the paradigm shift in planning was examined, specifically the master planning and participatory planning paradigms were reconsidered with regards to their significance in planning for regional integration. Also examined, as the object of planning that has implications for the development of the space economy, are the extended metropolitan region (EMR) and growth triangle (GT) and polycentric models of space economy found largely in Asia and Europe. As spatial models it is not clear how they relate with regional development theories, which sought the economic bases of integration, although lately contributions in new economic geography which developed in the 1990s tended to look more towards geographic space.

References

Acs ZJ, Varga A (2002) Geography, endogenous growth, and innovation. Int Reg Sci Rev 25 (1):132–148
Anas A (1980) Metrosim model. The state University, New York, Buffalo
Arrow KJ (1962) The economic implications of learning by doing. Rev Econ Stud 29(3):155–173
Batty M (2002) The emergency of cities' complexity and urban dynamics centre for advanced spatial analysis. University College, London, UK, pp 1–19

References

Barro RJ (1990) Government spending in a simple model of endogenous growth. J Polit Econ 98 (5):S103–S125

Boudeville JR (1966) Problems of regional economic planning. Edinburgh University Press, Edinburgh

Bretschger L (1999) Knowledge diffusion and the development of regions. Ann Regional Sci 33 (3):251–268

Burgess EW (1925) The growth of the city. Proc Am Soc Soc No 18:85–89p

Chirot D, Hall TD (1982) World-system theory. Ann Rev Sociol 8:81–106p

Christaller W (1933) 1966. *Central Places in Southern Germany* Translated by C. W. Baskin, Englewood Cliffs, Prentice-Hall, New Jersey

Chukuezi CO (2010) Urban informal sector and unemployment in Third World cities: the situation in Nigeria. Asian Soc Sci 6(8):133

Dawkins CJ (2003). Regional development theory: conceptual foundations, classic works, and recent developments. J Plann Lit 18(2):131–172

Dixon RJ, Thirlwall AP (1975) A model of regional growth rate differences on Kaldorian lines. Oxford Econ Papers 27, 2: 201–214 (cited in Dawkins, 2003:139)

Friedman M (1966) Regional development policy: a case study of Venezuela. MIT Press, Cambridge, MA

Friedmann J (1972) A general theory of polarized development. In Hansen NM (ed) Growth centres in regional economic development. Free Press, New York

Fujita M, Mori T (1998) On the dynamics of frontier economies: endogenous growth or the self-organization of a dissipative system? Ann Regional Sci 32(1):39–62p

Glasson J (1984) An introduction to regional planning: concepts, theory and practice (The built environment series), 2nd edn. Hutchinson Educational, London

Goldfrank WL (2000) Paradigm regained? The rules of Wallenstein's world-system method. J World-Syst Res 6(2):150–195p

Goodman R (1979) The last entrepreneurs: America's regional wars for jobs and dollars. Simon & Schuster, New York

Harris CD, Ullman EL (1945) The nature of cities. Ann Am Soc Polit Sci 242:7–17p

Holland S (1976) Capital versus the regions. Macmillan, London

Hoover EM, Fisher JL (1949) Research in regional economic growth. In: *Problems in the study of economic growth*, Universities-National Bureau Committee on Economic Research. National Bureau of Economic Research, New York (cited in Dawkins, 2003:140)

Hoyt H (1939) The structure and growth of residential neighborhoods in American cities. Federal Housing Administration, Washington, D.C

IDS (Institute of Development Studies) (2002) Learning Initiative on Strengthening Citizen Participation in Local Governance. An international Workshop on Participatory Planning Approaches for Local Governance held in Bandung, Indonesia from 20–27 January as part of LogoLink initiative. 2002

Isard W (1955) Hybrid land use representation. Available at: http://people.hofstra.edu/geotrans/eng/ch6en/conc6en/hybridlu.html. Accessed 2 Aug 2013

Isard W (1956) Location and space-economy. MIT Press, Cambridge, MA

Kaldor N (1970) The case for regional policies. Scottish J Polit Econ 17(3):337–348

Krugman P (1999) The role of geography in development. Int Regional Sci Rev 22(2):142–161

Landis J (1995) Imaging land use future: Applying the California urban futures model. J Am Plann Assoc 61:438–458

Losch A (1954) The economics of location. Yale University Press, New Haven, CT

Markusen A (1985) Profit cycles, oligopoly, and regional development. MIT Press, Cambridge, MA

Markusen, AR, Lee, Y-S, Sean D (eds) (1999) Second tier cities: rapid growth beyond the metropolis. University of Minnesota Press, Minneapolis

Martinez-Vela CA (2001) *World systems theory*. ESD 83

Massey D (1984) Spatial divisions of labour: social structures and the geography of production. Macmillan, London

Macleod S, Mcgee, TG (1996) The Singapore-Johore-Riau Growth triangle: an emerging extended metropolitan region. In Lo F, Yeung Y (eds) Emerging world cities in Pacific Asia. United Nations University Press, Tokyo/New York/Paris

Molotch H (1976) The city as growth machine: toward a political economy of place. Am J Sociol 82(2):309–332

North DC (1990) Institutions, institutional change, and economic performance. Cambridge University Press, Cambridge, UK

Ogbazi JU (1992) Comprehensive planning and master plan making. In: Mba HC, Ogbazi,JU, Efobi KO (eds) Principles and practice of urban and regional planning in Nigeria. Meklinks Publishers, Awka, Nigeria, 145p

Okosun AE (2013) Determinants of the nature and pattern of land use dynamics in Nigeria: a comparative analysis between Enugu and Benin City. PhD 2nd Seminar presentation. Department of urban and regional planning, University of Nigeria, Enugu Campus, Nigeria. 3rd March

Perloff HS, Edgar SD, Lampard EE, Muth RF (1960) Regions, resources, and economic growth. John Hopkins University Press, Baltimore, MD

Pettit CT (2002) Land use planning scenarios for urban growth: A case study of approach. (Unpublished PhD thesis, Centre for research into sustainable urban and regional futures, School of geography, planning and architecture, University of Queensland, Australia

Porter ME (1990) The competitive advantage of nations. Free Press, New York

Pred A (1977) City systems in advanced economies. University of California Press, Berkeley (cited in Dawkins, 2003: 141)

Rabinowitz P (2013) Participatory approaches to planning community interventions. Available at: http://ctb.ku.edu/. Accessed 15 October 2013

Ratcliffe J (1981) An introduction to town and country planning, 2edn. Copp Clark Pitman, London

Richardson HW (1973) Regional growth theory. Macmillan, London (cited in Dawkins, 2003: 147)

Richardson HW (1979) Aggregate efficiency and interregional equity. In: Folmer, H, Oosterhaven, J (eds) Spatial inequalities and regional development. MartinusNijhoff, Boston

Richardson HW (1981) National urban development strategies in developing countries. Urban Stud 18:267–283p

Romer PM (1986) Increasing returns and long-run growth. J Polit Econ 98(5):71–102

Rostow WW (1977) Regional change in the fifth Kondratieff upswing. In: Perry, DC, Watkins, AJ (eds) The rise of the sunbelt cities, Sage, Beverly Hills, CA

Schumpeter J (1947) Capitalism, socialism, and democracy. Harper, New York

Scott AJ (1992) The collective order of flexible production agglomerations: Lessons for local economic development policy and strategic choice. Econ Geogr 68(3):219–233

Solow RM (1956) A contribution to the theory of economic growth. Q J Econ 70(1):65–94

Steiber SR (1979) The world system and world trade: an empirical exploration of conceptual conflicts. Sociol Q 20(1):23–36

Swan TW (1956) Economic growth and capital accumulation. Econ Record 32(44):334–361

van Hamme G, Pion G (2012) The relevance of the world-system approach in the era of globalization of economic flows and networks. Geogr Ann Series B, Human Geogr 94(1):65–82

van Langenhove L, Costea A (2007) The relevance of regional integration. United Nations University Press, Tokyo/New York/Paris

Vernon R (1966) International investment and international trade in the product cycle. Q J Econ 80(2):190–207

Wallerstein IM (1976) The modern world system: capitalist agriculture and the origins of the European World economy in the 16th Century. Academic Press, New York, pp 229–233

Wallerstein I (2004) World-systems analysis: an introduction. Duke University Press, USA, p 128p

Weber A (1929) Theory of the location of industries. University of Chicago Press, Chicago

Chapter 6
Overview of Urban Planning Principles and Practice

Abstract The inception of participatory planning perspectives in Africa in the 1980s was a direct result of global trends linked with the activities of external assistance agencies. As neoliberal participatory planning perspective (planning without a plan), in which a market metaphor in development processes determines planning approach, the relevance of formal planning as a rational instrument to manage the distribution of activities in space diminishes in practice. Regardless, quantitative and qualitative research was used to establish the resilience of formal planning. Thus formal master planning remains a universal planning instrument and the idea of new master planning is underway amidst upsurge in neoliberal planning initiatives interpreted loosely as participatory planning perspective. It was found that spatial integration eludes neoliberal participatory planning initiatives in Africa.

Keywords Participatory · Master planning · Market · Africa · Capitalism · Imperialism

6.1 Dynamics of Paradigm Shift in Planning Perspective

There are three factors that influence the dynamics of paradigm shifts in planning. They are either global trends or the influence of northern-based development agencies or new approaches in response to particular issues in southern cities (Watson 2009: 179). Among the three factors, global trends are most overbearing and largely responsible for the insurgence of stakeholder participation in planning that is currently rocking planning perspectives worldwide. Reactions to this trend are insightful.

The assumption of a relatively homogeneous civil society with a common worldview, able to debate planning alternatives and reach sustained consensus that underpins the participatory theory has been challenged more generally and particularly in the context of Sub-Saharan Africa in the global south (see Watson 2002, 2006; Devas 2001). Although, as Watson (2009: 159) observed, 'there has been a tendency in the planning literature to assume a one-dimensional view of civil society and the role it might play in planning initiatives. The ideal of strong community-based organizations, willing to meet late into the night, debating planning ideas, may be achievable in certain parts of the world, but civil society does not always lend itself to this kind of activity'. In Africa, the Middle East and much of Asia, Bayat (cited in Watson 2009: 159) argues, "… social networks which extend beyond kinship and ethnicity remain largely casual, unstructured and paternalistic". In many parts of the world as well, Davis (2004) argues, civil society is being inspired more by popular religious movements (Islamist, and Christian or Pentecostal) than by organized demands for better infrastructure or shelter, given that efforts to secure the latter have so often failed. Overall the global south does not present a strong civil society that demands involvement in urban planning. The situation in Africa is such that inclusive participatory planning is more or less induced as a fallout of international development discourse.

The inception of participatory planning perspectives in Africa in the 1980s was a direct result of global trends linked with the activities of external assistance agencies. The participatory paradigm is more rhetorical than operational in spite of the mechanism of decentralizing planning administration to local government administrative levels. The effectiveness of this strategy had to contend with the inherent weakness of the local government system. Public participation and consensus, whereby the wishes of individual, small groups and the popularity of politicians shape urban destiny should be viewed seriously because it leaves room for unintended consequences of in-formalization most likely to outweigh the very elusive anticipated benefits. So far there has been an overreaction on the part of urban planners, leading to excessive participation and the neglect of actual physical planning.

6.2 Trends in Statutory Planning Perspective

Statutory master planning was and still is a universal planning instrument. Master plans prepared within the first half of twentieth century, especially by French architects under the influence of Le Corbusier, abound. Within the same period Japan experimented directly with imposed master planning and western urban forms in what were then its own colonies of Taiwan, Korea, China and Manchuria —see Hein (cited in Watson 2009: 174). Indian and Latin American cities' involvement is outstanding. Lately, in the 1980s, cities in China and East and South-east Asia, hitherto without institutionalized planning systems, adopted master planning amidst the contemplative scenario for new innovations in urban planning. Remarkably, Singapore and Hongkong within the Asian bloc have long-standing and successful experience with master planning. The new entrant China was formally rehabilitated with the City Planning Act of 1989, which set up a comprehensive urban planning system based on the production of master plans to guide the growth of China's burgeoning new cities (Friedmann 2005).

With the inception of colonial urbanization in the mid-19th century the instruments for urban design in Africa, specifically Anglophone Africa, were imported as direct products of the professional design tradition in urban planning in Europe, especially British town planning laws, the Town and Country Planning Act of 1947. A different approach somehow applied in Francophone Africa where colonial authorities fabricated planning instruments in situ as fortuitous circumstances permitted. Anyhow, the use of master planning instruments prevailed.

Since 2000 issues of environmental determinism, such as global warming, urban sustainability, climate change, etc., have propelled the trend to reconsider the tenets of master planning for efficient intervention. Already the idea of new master planning is underway in a few places, including China and in Singapore, where it takes the form of draft master planning. In both instances renewed master planning retains an intervention orientation and growth and land-use planning are its central elements. Its resilience is outstanding in the face of fierce criticism. Indeed, the problem is not with the instrument but with the instrumentality. Most of the criticisms, such as questioning its validity in a pluralistic society, restrictive, unrealistic, etc. are skewed reactions to human problems linked with spurious planning practices.

6.3 Trends in Spatial Planning Perspective

Since attention dimmed in terms of the master planning category of spatial planning, a renewed interest in spatial planning has found expression in guidelines to spatial frameworks within different spheres and levels of government. Harrison and Todes (2001) summarized reactions worldwide thus:

> The European Union (EU)… has prepared a European Spatial Development Perspective (ESDP) which provides a platform for negotiating the spatial allocation of resources across the continent. In the UK serious consideration is now being given to the preparation of a National Spatial Perspective (Alden, 1999; Shaw, 1999), whilst spatial frameworks have recently been prepared for Wales and Northern Ireland, with Scotland to follow suite (Lloyd and McCarthy, 2000). In the US, the current emphasis on 'growth management' and 'smart growth' is associated with a new wave of spatial planning at state, district and local scales, although there is still considerable variation between states (APA, 1999).

The EU encounter that led to the final version of ESDP in 1999, after ten years of continuous effort, is very illuminating. The ESDP is a shared vision for the European territory that arose primarily out of concern for the distortions and disparities in spatial development of EU territory. The intended operation is then meant to achieve a more integrated spatial development that will enhance the competitiveness of the territory in the global economy. Incidentally, the EU territory shares four regional perspectives in spatial planning. The challenge ahead was to integrate these perspectives and come up possibly with a unified European spatial planning perspective. How this can be done must find a methodology of accommodating diversities in the regional context of development. Hence in terms of the formulation of European spatial planning policy guidelines, the four spatial planning blocs held their bargaining positions to express their concerns.

The north-west perspectives spearheaded the collaborative process. It postulates the need for a formal planning competency that can lead to the inclusion of territorial cohesion as a shared competence in the Treaty establishing a Constitution for Europe (Faludi 2005). British perspectives were more concerned with the complex links between spatial planning and land-use planning. Nordic perspectives spotted the discursive nature of European spatial planning and Southern perspectives—apparently the missing link in the overall puzzle prefers to watch changes in planning practice. Janin and Faludi (2005: 211) synthesized these positions in their analysis of spatial planning as an experimental field for European governance.

Since ESDP, the configuration of spatial planning policy at a European level has manifested recourse to new policy processes, instruments and techniques (Giannakourou 1996: 608). The narrative that follows shows that the European context and the ESDP are beginning to be accepted as an important frame of reference in the production of Regional Planning Guidance. However, it appears that changes or innovations in governance occasioned by objectives of European integration generate ripples over a shared understanding of what European spatial planning actually means. Sure enough, there is still a long way to go, and European spatial planning needs a clearer technical definition of what it is about in order to make its usefulness and capacity as a proper tool of European integration more transparent (Janin 2003, 2005). They will have to find a way of fully incorporating the southern perspective whose planning tradition tends to lean more towards neoclassical perspectives as a practical option to achieve a more balanced and multi-centric system of cities.

The turn of events greatly influenced by ESDP compelled Dasí (cited in Nadin and Stead 2008: 40) to argue that the comprehensive integrated and regional economic planning styles are becoming more common, and, moreover, that this process is producing a "neo-comprehensive integrated planning approach". How this plays out in different scenarios depends on recent welfare reforms. Dutch spatial planning is tending towards a more liberal approach. It is now more difficult to categorize the planning system in England as dominated by the land-use regulation model (Nadin and Stead 2008: 44). Most of the changes derive from reformed planning legislation. The case of the reformed Planning Act of the 1970s, that decentralizes decision-making authority and promotes public participation in planning in Denmark and the Spatial Planning Act (2001) in Bulgaria are typical examples that identify are course to a comprehensive spatial planning approach. The 2004 Planning and Compulsory Purchase Act for EU countries introduced critical reforms for local planning. Planning here (ostensibly under the influence of ESDP) remains within the drawing board category and is essentially two-dimensional in emphasis.

The concept plan of Singapore is a peculiar approach that developed outside the influence of ESDP. It actually preceded ESDP and consistently maintained a basic physical approach to planning. Hence it remained within the realm of neoclassical master planning with a decentralization policy, enlarged regional reach and visioning capacities. This was integrated into a wider national policy of egalitarianism as Singapore evolved from a British colony to an independent city state (Eng 1992: 183).

Also not overtly committed to ESDP is the situation in Italy where a regulatory planning approach has traditionally been adopted (Janin 2003: 68). During the past decade Italy started to adopt strategic plans in response to a progressive shift of technical focus from city plans to urban policy and perhaps to the cities. In this very typical situation trends in planning indicate that many opportunities exists for integrating 'urbanism' traditions, regulatory requirements and the strategic dimensions of planning (Janin 2003: 66). The overall synthesis including the Singapore experience signals a gradual shift of emphasis to visual planning or more appropriately three-dimensional planning. At the horizon of this trend is creative urbanism, which is currently subsumed in contemporary new urbanism. Creative urbanism emphasizes variety—the focus is not on technicalities but on visual appeal, cultural significance and aesthetic innovations that are based on the theory of related areas (Tunnard 1951: 234). This approach to planning perceives the city as a worthwhile expression of culture and art. In its operations it tends to integrate the past with the present while planning for the future.

In Africa two reasons are responsible for the renewed interest in spatial planning in the mid-1990s. First was the bandwagon effect of global trends in relating spatial planning to infrastructure and second, the severe criticism of master planning allegedly for its inability to shape spatial changes in cities (especially those in developing countries) (Todes 2012: 158). For most African countries the first reason is more pronounced, except in the rare case of South Africa where the rejection of master planning left a vacuum in spatial planning that elicited attention in the first instance and then followed by introducing a trend. The South African experience is sufficiently visible in literature.

Within roughly four decades of continuous inquiry South Africa experimented with roughly seven alternative approaches to spatial planning, each with an average lifespan of 5–6 years. Yet the highly elusive search for an appropriate paradigm for spatial planning seems not to be on hand. However, some milestones have been reached in three directions: firstly, linking spatial planning with infrastructure development or sectoral planning, secondly confirming that a recourse to neoliberal ideology had the effect of politicizing planning beyond the traditional levels of politics in planning, and thirdly, realizing the strategic need for statutory planning, strong leadership, and detailed plans with which to direct development as prerequisites for effective planning intervention. These are hard facts, ordinarily in the rhetoric of neoliberal planning, and will not be easy to accept.

From the foregoing events in the EU and perhaps the ASEAN bloc, it seems that the trend is very pragmatic, focused on the problem and committed to a vision. In the developing countries, spatial planning was marginalized by structural adjustment programmes in the 1980s but from the events in South Africa there do seem to be indications of a restored role for spatial planning as part of the broader shift towards integrated development planning (Harrison and Todes 2001: 65).

6.4 Trends in Spatial (Urban) Planning Initiatives

Since the 1980s spatial planning initiatives have altered, following a paradigm shift from formal to participatory planning. Each country responds according to its economic fundamentals and its national vision for development. As participatory principles dominate the planning landscape, lots of experimentation are underway and innovative planning systems and new structures continue to emerge. The growth of normative frameworks is evident alongside the development of subjective institutional models some of which have coalesced under the auspices of Cities Alliance to generate a Cities Development Strategy (CDS). Many countries, both in the developed and developing countries, tend to imbibe the CDS culture which indicates economic growth, participatory planning and spatial form qualities driven by the World Bank, UN-Habitat and the Japanese government respectively. The different qualities elicit different approaches to CDS and the UMP and EPM approaches are commonplace. In Africa the UMP approach is very popular for poverty alleviation, especially in Francophone African countries, the bulk of which is located in West Africa. Anglophone African countries tend to be more inclined towards normative initiatives however without prejudice to CDS set of guidelines using EPM approach although EPM is linked more with sustainable cities programme (SCP). In the past four decades these alternative sources of initiatives have influenced participatory planning in Africa.

6.4.1 Review of Sustainable Cities Programme (in Nigeria)

The SCP is a capacity building instrument that offers a practical response to the universal search for sustainable development. SCP is a joint facility of UNCHS and UNEP conceived to package and apply environmental technologies and know-how by strengthening local capacities to address urban environmental problems by municipalities and local partners with multi- and bilateral external support (UNCHS, UNEP, UNDP, World Bank, WHO, ILO, Germany, France, Denmark, Sweden, Japan, Canada, the Netherlands, USA). Its stated key features are: broad-based stakeholders' involvement rather than master planning; bottom-up problem-solving rather than top-down decision-making; mobilization of local resources; framework for coordination of external support and environmental awareness (UNCHS 1996).

In the words of UNCHS (1996: 6):

> The SCP is pre-eminently a locally-focused program, in which national, regional, and global support is built up from activities and experiences at city level. The SCP provides a framework for linking local actions and innovations to activities at the national, regional, and global levels, through which the lessons learned in individual city experiences can be shared, analyzed, generalized, and discussed widely. This operational link serves to make global strategies more responsive to local needs and opportunities and, conversely, helps to implement global strategies and agreements at the local level. As a global program, the SCP

6.4 Trends in Spatial (Urban) Planning Initiatives

promotes the sharing of know-how among cities in different regions of the world. As an inter-agency effort, SCP mobilizes, packages and applies technical and financial resources from diverse sources.

At its four operational levels SCP has specific aims: at city level to strengthen local environmental capacity in city demonstration projects; at country level to address national issues arising from city level experience and to promote replication strategies for cities throughout the country; at regional level to pursue four-pronged objectives (to facilitate the systematic collection, assessment, sharing and dissemination of information; to share operational lessons of experience in urban environmental management; to promote the pooling of scientific and technical resources; and to support the creation of joint agendas and coordinated programmes of action); and at global level where it is driven by the needs and lessons of local demonstrations to catalyze capacities of the UN system for local, national, and regional capacity building in urban environmental management (UNCHS 1996). Typical issues addressed by SCP include water resources management, air pollution and urban transport, environmental health risks, flooding and drainage, unstable slopes, industrial risks, solid waste management, urban agriculture, recreation and tourism resources etc (UNCHS 1996). These are environmental management issues and the aim of SCP is to incorporate them into urban development decision-making.

For purposes of achieving these objectives new approaches to urban governance and administration were sought because master planning approach (MPA) was considered inappropriate. The environmental planning and management (EPM) option was preferred and adopted as an articulated analytical framework and logical structure which facilitates better understanding of the dynamism of urban development and environmental issues and helps in evolving convincing guidelines (or strategies) for intervention (UNCHS 1996). The EPM framework is meant to promote a stakeholder participatory approach to decision-making in urban environmental management and provides tools and processes which allow cities in prioritizing environmental issues, decision-making and implementation (UNCHS 1996). EPM is a process designed to serve as technical bases for SCP.

Based on the principle of cross-sectoral interaction and inter-organizational collaboration to ensure sustainable development the EPM as a process is meant to implement agreed strategies programmed to solve identified environmental issues through coordinated public and private actions. The EPM process has four elements; namely, improving environmental information and expertise, improving environmental strategies and decision-making, improving effective implementation of environmental strategies and institutionalizing EPM. From these elements Ayorinde and Asamu (2000) derived a nine-stage EPM process thus: project documentation; environmental profile; city consultation; issue-specific working group for issue specific strategy; environmental management plan; action plans; project implementation; replication of the process; and monitoring/review. Institutionalizing EPM is an integral part of all the stages except for the initial and second stages. All of the stages have a set of specific assignments set out to be concluded within a maximum of 27 months.

Having reviewed the SCP/EPM process certain features are apparent and they are:

i. EPM is a global network of planning methodology proposed for the SCP,
ii. SCP/EPM process is an incremental planning strategy. It is not futuristic,
iii. The process has inbuilt stakeholders participation system,
iv. It adopts normative controls rather than traditional planning controls,
v. It is non-spatial in content, and
vi. It is project based.

Given these observations it is not clear how SCP/EPM process applies in modern urban planning. This uncertainty elicits some pertinent questions such as these:

i. Does EPM set planning guides such as the extent of planning area, plan period, etc.?
ii. Are there provisions for linkages with sustainable urbanism?
iii. Who coordinates the process at city level?
iv. What is the relationship between Environmental Action Plan (EAP) of EPM process and sustainable urbanism?, and
v. Apart from being global in scope how do SCP/EPM operations differ from planning practice in the pre-professional design tradition period?

As we consider these questions we can summarily conclude that the vision of SCP/EPM is to integrate stakeholders' participation into environmental management and its mission is to prepare EAP through interactive forum; while its aim is to set up a global network of planning practice. We shall endeavour to establish its links and relationships with sustainable urbanism as we examine its analytical framework.

It is claimed in literature that 20–30 cities now subscribe to the SCP/EPM scheme since its inception in 1996. Most of them presumably are amongst those identified for UNCHS pilot projects. The UNCHS pilot scheme is meant to be a global scheme but remarkably all of the cities numbering twelve are located in developing countries in Asia and Africa. Maybe developed countries are not favourably disposed to the idea or they are in it to provide bilateral and technical assistance.

In 1999 the UNCHS conducted an evaluation of SCP projects in six cities in the African region. The six cities were: Accra, Ghana; Dakar, Senegal; Dar es Salaam, Tanzania; Ibadan, Nigeria; Ismailia, Egypt; and Lusaka, Zambia. Apart from reviewing and assessing the utilization and impact of the donor support the primary aims of the overall evaluation were to assess the progress of each project and to extract lessons of experience which can help improve overall programme design and implementation of similar projects in the future.

The evaluation identified the rationale for engagement in SCP/EPM projects as follows: industrial risk and pollution in industrial neighbourhood (Dakar); rejection of MPA (Dar es Salaam); deteriorating urban environment with drastic decline in the ability of local government to deal with the situation (Ibadan); environmental

6.4 Trends in Spatial (Urban) Planning Initiatives

consequences of the city's expansion (Ismailia); and rapid growth of unplanned settlements on the outskirts of Lusaka. The assessment of the local context indicated that the political, cultural and institutional contexts are not always favourable for the promotion, adaptation and integration of the SCP process (as it was the case for MPA) due to frequent changes in local government leadership and rapid turn-over of professional staff officers in local (and state) government. For Lusaka financial shortage and lack of qualified manpower was the problem. In Ismailia apprehensiveness was the problem.

The SCP/EPM scheme was project specific and depended largely on external support although the mobilization of local resources was reported for Ibadan. The assessment of capacity building and the importance of the process in leading to attitudinal and institutional changes yielded negative result. The concept did not resonate with local institutions. It was discovered that the period of 2–3 to bring about fundamental changes and long lasting impact is too small. Longer time is required thus discrediting the short-term hypothesis of SCP/EPM process. The short time frame lead to great pressure on project management, elicited the expectation of quick results, and necessitated rescheduling of projects with the implication of securing agreement from funding agencies to extend the project as it was the case in Accra and Ibadan. Furthermore monitoring and assessment of project progress was virtually non-existent in all six projects. These are the same ailments that suffocated MPA including general low level of public awareness and understanding of the SCP/EPM concept, except perhaps in Ismailia where awareness is appreciable. Also in Dar es Salaam, Ibadan and Ismailia there are indications that the SCP project is making some impact on professional training given to planners, geographers and others. In a related assessment Schoonraad (2000) states:

> The emphasis on public participation has brought the needs of individual communities to the forefront at the expense of the common good. It has also focused on short-term needs as opposed to long term sustainability goals. In drafting of integrated development plans for Pretoria, it became clear that people were only concerned with their own areas, and had little concern for the effects on the rest of the city

In practical terms the impact on professional training ultimately strengthens the knack to remodel substantive development concepts. This attitude is common place in Africa and already the remodelling of the SCP/EPM concept is manifest in spite of the relative short time frame of its application. In Dar es Salaam a new planning approach tagged Strategic Urban Development Plan (SUDP) is being developed. In Nigeria independent sources are already proposing alternative approaches in the absence of authentic analytical base: Ayorinde and Asamu (2000) suggested Integrated Physical Planning Approach (IPPP); Jiriko (2004) postulated several options including Strategic Environmental Planning and Management (SEPM); Ilesanmi (1998) suggested Community-Based Approach (CBA) and so on. And true to SCP/EPM tradition, none of these proposals has a clear bearing on sustainable urbanism. They are all project oriented and will end up with the preparation of Environmental Action Plan (EAP). This trend of concept remodelling which feeds inordinately on the escapist attitude of planners in tackling problems of plan implementation is likely to continue ad infinitum unless we learn how to be pragmatic. At the moment concerted action in robust application of development concepts seldom prevails over rhetoric and indefinite search for save haven.

Nonetheless the evaluation exercise indicates that progress was made on replication (Dar es Salaam); resource mobilization (Ismailia); establishing framework for coordination (Dar es Salaam); better information flow (Lusaka); and establishing framework for coordination and resource mobilization (Lusaka). A summary of general assessment of all six projects is given in Table 6.1.

The evaluation established some preconditions for successful SCP project design and implementation. They are good timeframe, quality project leadership, competent and dynamic working groups and more importantly local ownership and political commitment which manifest itself in budgetary allocations to initiatives generated through the process. Ironically the same context is required for successful MPA application but unfortunately rather than address these contextual issues in the case of MPA and as it is now the case with SCP/EPM, attention is rather focused on sourcing new concepts that will probably adapt to the prevailing context. This is not acceptable because the prevailing context is not desirable. Furthermore the same limitations that grounded MPA are at work on SCP/EPM. The findings of the

Table 6.1 A summary of general assessment of SCP initiatives

EPM Elements	Overall assessment by the evaluation team
Improving environmental information and expertise	Did reasonably well; overview info well served by the profile; stakeholders effectively involved through city consultation and working groups; priorities set at the city consultation
Improving environmental strategies and decision-making	Did somewhat less well; best results in clarifying policy options; least satisfactory with respect to considering implementation options and resources; fairly good success in building consensus; mixed results in coordinating environmental and other development strategies with better success in some projects (Dar es Salaam, Ismailia) than in other (Ibadan, Dakar)
Improving effective implementation of environmental strategies	Mixed results; limited success in applying full range of implementation capabilities, with general reliance on capital investments; well done in agreeing on action plans; less well done in developing packages of mutually supportive interventions; weak in reconfirming, politically support, critically weak in Accra and Ibadan
Institutionalizing EPM	Disappointing; some success in strengthening system wide capacities for EPM. Signs of progress in accepting broad-based participatory approaches with difficulty in moving from acceptance to incorporation; least progress in institutionalizing cross-sectoral and inter-organizational cooperation; little success in monitoring, evaluating and adjusting the EPM system

Source UHCHS (1999)

6.4 Trends in Spatial (Urban) Planning Initiatives

evaluation exercise indicate that project design features were either inappropriate or subject to misinterpretation and/or misapplication. Besides, financial management requires urgent attention because it impacts on project timeframe and successful completion of projects. The need for internal monitoring systems was identified as an integral element of project management. In the words of the UNCHS (1999) report: 'there has been little or no provision in the project design for setting up systems for monitoring the progress of the SCP process itself.' To this end the report observed that city consultation should be encouraged to proceed along positive track. The evaluation report also highlighted the need for attention to focus on demonstration projects because demonstration projects promote the ideals of SCP/EPM. Hence there is need to generate "fast track" community-based demonstration projects and organize their implementation in a participatory but expeditious way. The system must also be very carefully designed so that demonstration project money does not reinforce the common misperception that the SCP project is able to provide capital funding for bankable projects.

In Nigeria, within ten years of experimentation of the twin concept, complacency is already manifest. Lessons from Sustainable Ibadan Project (SIP) that highlight the adverse impact of these conditions have been reviewed earlier in the context of SCP initiative in five other African cities. Remarkably the two other SCP initiative in Nigeria namely, Sustainable Kano Project (SKP) and Sustainable Enugu Project (SEP) are underway without significant reference to SIP experience. In SEP the GIS equipment and urban observatories are barely functional as intended. Inter-departmental in-fighting, lack of cooperation, insincerity and political influence are all common place problems because SEP's link and relationships with existing planning machinery is not explicit (Okeke 2005: 87). The researcher once attended a meeting of SEP stakeholders' forum and confirmed the futility of adopting stakeholders' forum as an instrument to encourage public participation and liberalize planning decisions. From the proceedings of the meeting it is obvious that such omnibus crowd do not have the competence to deal with issues of mainstream planning. SKP is no exception and by the way the euphoria about SCP initiative in Nigeria is practically dead (at least in Enugu) without making any serious contribution towards urban management. Ironically, in spite of the absence of concrete evidence and true to the hypocrisy that mark the reactions of some urban scholars in Nigeria Ayorinde and Asamu (2000) claimed the replication of SIP in SKP and SEP. He also proclaimed success story for SIP and SCP/EPM in general in Nigeria. Meanwhile the premise on which he made his proclamation is the basic information he provided on SIP working groups and this is not enough to justify his claim. Many of this kind of misleading campaign abound to service regional interests.

However having reviewed theoretical and analytical frameworks for SCP/EPM it appears there is a kind of disconnect between the new concept and sustainable urbanism. At the outset the bias against spatial planning launched SCP/EPM. So SCP/EPM is a non-spatial concept best suited for project development rather than urban planning. It is a process-oriented instrument without professional bias and modelled after the muddling-through system of incremental planning. It is not clear how an instrument with these qualities can be useful to sustainable urbanism. As a control instrument it has no locus whatsoever in urban design, its administration is prone to mediocrity due to its populist orientation, and it is susceptible to encourage sprawl given its disposition to disjointed approach to project development. The vagaries of incremental planning and its potentials to create sprawl are clear. In the SCP/EPM dispensation the perception of the dynamics of urban systems is too simplistic and relegates to the background the intricate and integrative exercise that is required to direct the destiny of urban form. Planning seeks to influence and not necessarily to contain or conform to future trends. Hence it encourages the design of futuristic end products. Theoretically SCP/EPM and its end product EAP are far from meeting these conceptual planning principles and standards. Maybe SCP/EPM could play a complementary role in planning process in terms of procedure but certainly it cannot assume the position of alternative urban planning concept, at least not in Africa where cities are yet to have defined spatial framework.

The wave of change from MPA to SCP/EPM in developing countries is unfortunate considering the inadequacy of SCP to address urban instability which characterizes urbanization in these countries. The analytical framework confirmed that the SCP/EPM is not free from the contextual predicaments of MPA so where is the rationale for its application. In fact SCP/EPM presents a worst case scenario because of its reliance on external funding. We are all aware that external funding always indicates repressive tendencies. Otherwise how do we explain the control of the review of timeframe for SCP/EPM projects? Powerful multi- and bilateral external support use financial instrument to manipulate the SCP/EPM process and to entrench it as a planning approach. The dilemma this has introduced into planning practice, particularly in African countries, is very disturbing. Most countries are turn between SCP/EPM and MPA. This should not be the case because in spite of the acclaimed global reach of this SCP/EPM campaign developed countries are not experiencing this dilemma in their planning practice. Planning practice in developed countries is defined in favour of mainstream planning and oriented towards sustainable urbanism. All other forms of planning are subjects to this macro framework of urban planning. The relevance of the tenets of master planning is still fundamental in their effort to manage urban growth in their region. The African sub region should not jettison master planning without weighing the pros and cons empirically.

6.5 Trends in Planning Practice: A Case Study of the Resilience of Formal Planning Practice

The impression is strong in planning literature that neoliberal planning initiatives do not have the capacity to deliver integration in development processes. At least this is reasonably true in the African context. A disconnect between neoliberal planning theory and spatial modelling of space economy is also in contention as a result of the relegation of formal planning principles and the disregard for urban growth management instruments in neoliberal planning theory. Regardless of the overarching influence of informality in the neoliberal planning dispensation, there are clear indications of the resilience of formal planning practice. A contrary impression is tested as null hypothesis in a case study. The methodology and preliminary literature of the case study is hereunder presented.

6.5.1 Methodology

The case study, which was conducted within 2010–2015 period at North West University (NWU), tested the null hypothesis: Form-based planning attributes are not significantly resilient in the perception of planning initiatives in Africa (as measured by the nested perception of related variables in the theoretical framework of the IPD initiative in South Africa).

The research methodology as shown in Fig. 6.1 identified study areas for different study activities that seek spatiophysical bases for integration. For purposes of desktop studies of planning initiatives four countries were identified including Mali, Egypt. Tanzania, and South Africa. Respectively the initiatives identified for study include CDS, "Shorouk":, O&OD, and IDP initiatives. For purposes of desktop country profile studies the following countries were identified: DRC; Angola; Mali; Egypt; Senegal; Kenya; Nigeria; South Africa; Tanzania; and Ethiopia. The outcome of desktop perception studies identified IDP initiative in South Africa for

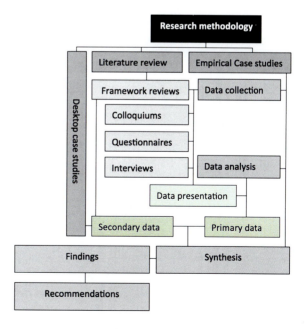

Fig. 6.1 Research methodology

empirical studies. For this purpose Tlokwe local municipality was identified as study area and Matlosana and Rustenburg local municipalities were chosen as location controls. The empirical studies engaged planning questionnaires and personal interviews as instruments for primary data collection. Four categories of respondents were investigated including: academics, politicians, administrators, and consultants.

The research methodology involved six sets of analysis, four for desktop studies and two for empirical studies. The Multi Criteria Analysis (MCA) instrument was engaged followed up with SWOT analysis and own assessment using personalized **4As** analytical template. The analytical approach is substantive not classic as in world systems analysis. Therefore the annals theory and dependency theory are fundamental elements in the value system of analyzing trends. Through the synthesis of descriptive statistics resulting from analysis, contribution is made to new knowledge and a new planning paradigm is theorized.

6.5.2 Preliminary Literature (Theoretical and Analytical Frameworks)

The structure of literature review is outlined in Fig. 6.2.

Literature study is categorized in two parts, namely related and relevant literature. Related literature focused on three elements: urbanization, urban growth, and the informal sector. The purpose of the studies was to understand the dynamics of these elements and how they are managed in the best interest of African economy. For urbanization studies a historical perspective was adopted starting from the cradle of African civilization to trends in contemporary urbanization. Patterns of urban growth and change in Africa were looked into and remarkably from a spatial perspective and not necessarily from demographic trend. The studies on informal sector focused on the informalization of human systems and the attributes of the sector as a determinant factor of population mobility that impact the distribution of people and activities in space.

The informal sector is a curious phenomenon presumably oversighted in contemporary planning. Thus it elicited special attention in the diagnosis of related literature. The insight gained indicates that the informal sector plays a central role in the bifurcation of space economy in Africa. It does so by encouraging spatially segregated dualist market economy in the economic landscape of urban regions. This works to encourage extroverted urban economy in the region in which local subsistent rural economy is disconnected. It is argued that the informal sector in its present survivalist characterization will not be able to harness optimal resource utilization in the urban region. It is also not clear if the sector possess requisite integrative qualities that will encourage functional flow of economic activities within the urban region. In other words there is no indication in literature that the functioning of the informal sector has the tendency to integrate the urban and rural economy.

6.5 Trends in Planning Practice: A Case Study …

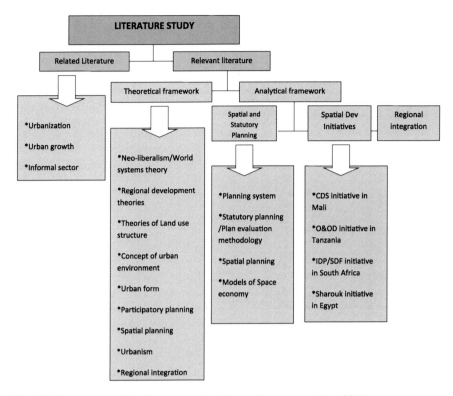

Fig. 6.2 Components of the literature review. *Source* Own construction (2011)

Inversely then, does the informal sector discourage spatial integration of space economy in the urban region? To what extent is the survivalist informal sector and dependency correlated? These are critical questions that should guide the optimism for informal sector operations. Onyebueke (2011) assertion that the African nations should embrace the challenge of in-formalization and chart her destiny courageously is neither here or there. The informal sector is implicated in the management of the space economy hence its links and relationships with the resilience of urban productivity decline in the sub-region and the unacceptable patterns of growth in African cities. Ultimately the tolerance of informality is dicey and attempts to regularize it should be treated with caution. This is why the recourse to informal planning begs the question and indicates a wrong diagnosis of the dynamics in the unstable form of urban regions in Africa. Overall the development of the informal sector is a big challenge to spatial planning because it threatens integration which is the central element of planning. With increasing urbanization the threat gains momentum and its growth is not unconnected with changes in development ideology which is influenced by global forces.

Overall the study of related literature underpinned the transition in the development surface of Africa from trends in its traditional civilization. This is marked by the inception of imperial space economy and the development of dependent urbanization. The spatial attributes of these changes is sprawl and slum development with attendant productivity decline. The resultant informal sector provided the rationale for neoliberal planning following transition to neoliberalism. In the circumstance informal planning thought to work against integration assumed prominence. This has the effect of deepening the bifurcation of space economy by means of encouraging disconnects.

On the other hand, the study of relevant literature which is in two parts reviewed theoretical and analytical frameworks of spatial planning. The concept of neoliberalism and neoliberal theory was examined with intent to establish the ideological perspective of neoliberalism. Then the evolution of regional development theories was examined from its outset in early 20th century until the current neoclassical theories that is epitomized in new economic geography theories. Through the investigations attention focused on the progression of contributions and their links with spatiophysical perspective. The choice of theories to be investigated stood on the existing protocol established by Dawkins (2003) in his overview of literature on regional economic growth. Dawkins (2003: 132) in his words "reviewed seminal works and comprehensive overviews of the most important theoretical concepts". This discuss accepts his approach and adopts the same set of regional theories for investigation however from a different intellectual perspective. The investigation earlier presented summarized contributions towards the development of regional development theories up to the development of neoclassical theories since 1990s (see Krugman 1999).

The study of urban planning theories focused attention in three directions: first, expounding the concepts of urban environment and urban form, the region and regional integration in their full ramifications highlighting attributes such as formal, economic, functional, planning region concepts that comply with contemporary understanding but seldom put into consideration in the rhetoric of planning. Inevitably classic theories of urban land use structure was revisited. The last known input in these set of theories was indeed in mid 1950s with the contribution of Isard (1955) hybrid theory. Second, the evolution in spatial planning was examined especially the impact of new perspective in planning occasioned by neoliberalism which spotlights informality and participatory planning. The dynamics this incidence generated in planning paradigm and institutional changes were reviewed relative to their significance in planning for regional integration. Hence models of space economy in Asia and Europe were examined. And third, design-oriented approach to planned development was examined in the context of the evolution in urbanism.

Overall, trends in the theoretical framework of planning was found to be under severe influence of informality. This trend received impetus from neoliberalism. Within the period investigated starting from 1960s, neoliberal planning theory developed to compliment trends in the development of regional development theories which were found to seek economic bases for regional integration and not spatiophysical bases although more recent contributions in new economic geography tended to look forwards geographic space. The growth of classic urban land use theories stagnated in the 1980–1990 period and that gave neoliberal planning theory the leverage to undermine land use planning. In effect urban from, urbanism and modelling of space economy lost favour. The only appearance of spatial integration was found in the compact city concept in reaction to humanistic interventions and environmental determinism that threatened sustainability. Otherwise there is no link between the development of regional development theories and spatial models of space economy. The urban structure theories had little to do with growth management instruments. Therefore theoretically there exist disconnect between the interacting elements. There is a yawning gap between theories and spatial modelling and informality consolidates this trend.

The analytical framework of planning was also examined in three directions: first, the planning system in Africa was examined vis-à-vis spatial and statutory planning structures and operations including plan evaluation methodology, its links and relations with plan implementation. To this end the evolution of spatial planning in South Africa was reviewed as reference point but the pattern of space economy was reviewed holistically and its evolution traced historically. Second, spatial development initiative in select countries from the four cardinal regions in Africa were assessed on their performance levels as participatory planning instrument. And third, NEPAD and NEPAD cities initiatives were reviewed as instruments for regional integration in Africa. Hence attention focused on reactions in the planning system vis-à-vis planning initiatives, perspectives and frameworks. The direction of reactions that determine the dynamics in the planning system is critical. To determine if the direction of reactions conforms to the disregard for spatial integration noticed in the theoretical framework or works to redress it is imperative. Hence the changes in the nature of planning and its application including the value system for qualitative planning were reviewed. The repositioning of planning rationality caused by these changes indicates for this research the direction of reactions.

Overall, the review of the analytical framework revealed that neoliberalism and informality made some inroads in planning perspective. Generally, transition from master planning to strategic planning perspective was noticed. The Anglophone bloc was found to be more forthcoming with the transition than the Francophone bloc. The entrance of participatory planning perspective is partly associated with plan evaluation methodology which links heavily with plan implementation. Most statutory plans were not implemented and this was considered a failure associated with the planning system. This philosophy provided the background to jettison master planning. The entrance of strategic planning favoured pro-poor planning and inclusiveness, a wisdom drawn from neoliberal foundation to accommodate growing informality and participatory planning. Europe reacted with the inception of European Spatial Development Perspective (ESDP) designed to address regional integration through territorial planning, America and Association of South East Asian Nations (ASEAN) remained focused on their growth vision and did not permit the new wave to becloud their commitment to urban design. In Africa most of the planning initiatives took the nature of strategic planning; thus pulling out of statutory planning under the influence of paradigm shift diffusion mainly from Europe and in practical terms the activities of UN-Habitat and other development partners.

It is concrete at least in the case of Africa that the direction of reactions in the change of attributes of elements in the planning system is responsive to the trend identified in the theoretical framework of planning. In other words the changes are not focused on integration. This draws from the increasingly disadvantageous position of planning rationality in managing the space economy. The combined influence of growing informality, informal spatial planning that encourage broad guideline plans, and sectoral planning work to enhance the stakes of market force in determining the development of the space economy. This is given expression in infrastructure development planning epitomized in Resource-based African Industrial and Development Strategy (RAIDS) initiatives. The impression is however given in the new concepts of planning initiatives that integration is a focal element. The RAIDS initiative sought that impression with its affiliation to Spatial Development Initiative (SDI) and development corridor strategy but RAIDS framework is more or less spatial distribution of project planning. Regardless the focal element is indeed on growth from economic perspective.

The analysis of literature summarily indicates lots of dynamism in the system as informality gives impetus. The new planning paradigm introduced by neoliberal planning seems to be gaining some foothold but it is not delivering integration because conceptually it is not form-based and economic and spatial planning are yet to integrate. However the new theoretical framework, which is identified to develop since 1960, is found to influence the conception of planning initiatives in Africa. Amongst the four African planning initiatives examined the Integrated Development Planning (IDP) in South Africa seem to be more compliant to the new theoretical framework.

However a critical change occurred in the meta-theory of planning by the end of the 1990s. Neoliberal institutions, especially the World Bank, admitted changes in orientation towards the role of the state in the development process. The new position encapsulated in the post-Washington consensus indicated the demise of the state-market dichotomy and the rise of a debate that was not concerned with state intervention per se but with the form and extent of that intervention and with building the capacity of the state to match its development tasks (Tawfik 2008). Regardless the direction of the debate, the alienation of the state from planning stands a chance to be revisited.

6.6 Summary of Basic Scenario of Planning Practice

The elucidation of the basic scenario is compelled to adopt a historical perspective for purposes of putting current trends in planning in their proper context and in the process exposing its real meaning and implications, rather than enrolling in the evaluation of spatial planning in Africa, based on compliance with new perspective

6.6 Summary of Basic Scenario of Planning Practice

of planning established with neoliberalism serving as intellectual instrument. The institutional school of thought postulates that current new perspectives in planning are primarily focused on enhancing planning implementation through participatory planning processes. This trend of thought has the effect of reducing the forces that influence paradigm change to local conditions hence in the rhetoric of paradigm change, global influences are seldom mentioned. But it goes without saying that for an overarching hegemonic phenomenon such as neoliberal planning to manifest global influence is inevitable.

Neoliberalism emanates from capitalism. The African continent encountered capitalism after a long sojourn with imperialism in the mid-19th century following contact with western civilization. This experience had implications on African space economy which are characterized by the emergence of colonial towns. The geography of the African economy was rewritten from what it was in the mercantilist period when the African continent controlled its destiny through local institutions and authority. Then around the 15th century African empires and kingdoms grew and were expressed in the culture of cities. Imperialism halted this trend first by reworking the mindset from nation-building to private sector development and the instrument for planning started its gradual transition from rational thinking which informed the then popular design tradition in planning. At the inception of the professional design tradition in the late 20th century, a master planning instrument was introduced and proved to be a very effective instrument to control the use of space. Its use was, however, tuned to deliver an imperial space economy under colonial authority.

After the independence decade (1950–1960) in Africa, nationalist governments inherited the master planning instrument but the challenge of capitalism, which included the influence of market forces in land-use management, proved difficult to disregard. The consideration of culture, value system and environmental factors in planning gradually morphed into considerations of funding mechanisms. Funding is an instrument of control in the neocolonial dispensation. Most African economies have issues with funding and this compelled them to cooperate with external development partners apart from having visionary leadership problems. By the 1970s market force considerations outplayed master planning on the heels of structural economic adjustment programmes occasioned by global economic recessions. Incipient project development syndromes sustained planning vacuums and informality gained a foothold as the common good lost impetus.

Market force is an instrument for managing global economy and the world system is insistent in making it an instrument for managing land use rather than planning rationality. Meanwhile African countries among other developing economies are in a disadvantaged position to engage in the competition market force encourages and they have little defence against the impact it exerts on their space economy. If nothing else, it erodes their control of the space economy. The dilemma of African countries in managing their space economy deepened as neoliberal economic policies emerged in the 1980s to initiate the era of neoliberalism, a concept that provides tacit support for market force consolidation. Neoliberalism arrived with new planning perspectives referred to as neoliberal planning and interpreted loosely as a participatory planning perspective. The new theoretical framework for spatial planning that issued from the neoliberal background sought two imperatives that are tangential to enhance the stakes of market force and they are: eliminating form-based planning and encouraging a broad-based guideline framework as planning instrument. Hence compliance with these standards should be seen for what they represent—instruments in the hands of neoliberals to usurp the authority of national governments to manage the development of their space economies. In other words, compliance in the circumstances that compromises the imperatives mentioned earlier cannot be seen as a virtue.

It was found that spatial integration eluded most planning initiatives in Africa in spite of positive compliance found across the board between principles and practice, yet the so-called plan implementation that informed the experimentation with neoliberal planning was at best marginally achieved. The inability to secure integration implies in practical terms that infrastructure provision is no longer subject to planning control, thus disproving the notion that form-based planning is not necessary for managing spatial (urban and regional) integration. This is why South Africa needs to be more tactful in pursuing a compliance that has the effect of installing market forces in urban planning because it is not a strategic approach to redress spatial distortions. The lessons of apartheid planning lie not in removing planning control but in redressing its racial content and realizing that master planning is a feasible and practical instrument to achieve common good or public interest. Compliance with paradigm shifts in neoliberal terms is a threat that removes the poor and weak at individual and even at country and continental levels from the purview of planning. The extent to which this threat is realized in conceiving planning initiatives is not clear; however, the weak disposition of planning initiatives is worrisome and appears to be directly related to compliance. This makes it vulnerable in the context of the syndrome of the market metaphor. The need to comply is most likely factored on remaining relevant in the world system rather than acceptance as the resilience of local planning practice indicates. However, it is consoling to note that the rhetoric of planning is changing faster than practice, as measured by the perceived resilience of formal planning. This leads to the proposition that master planning, however controversial, has not outlived its usefulness and relevance in urban planning in Africa.

The challenge ahead of a new contribution is to suggest a development paradigm that can achieve integration without compromising participation in planning. The practical problem with this challenge is the integration of economic and spatial planning to achieve growth and simultaneously shape the city. This requires the resolution of the interface between political and spatial analysis in development planning. The rider in this mission is to manage the role of informality and determine the place of the market metaphor fronted by neoliberalism. The best approach, perhaps, is to engage the market metaphor in the use of space in spatial planning initiatives. The role that formal planning principles play in this regard is critical and elicits the test of a research hypothesis which seeks to establish the status of the resilience of formal planning principles and practice.

6.7 Conclusion

Long before the incidence of neoliberal participatory planning, a master planning paradigm had existed. Criticism against master planning came in phases and metamorphosed into attempts to reinvent planning, which neoliberalism introduced. Developed countries, with integrated development plans, relied heavily on master planning principles currently transformed into urban design principles which complement planning which arguably has not moved significantly away in practice from formal planning Otherwise, recent legislation in some developed countries such as The Netherlands that excludes poor people from certain sections of the city, contravenes inclusive planning driven by participatory planning in the current dispensation of neoliberal planning theory. If countries that have integrated spatial systems still rely on formal planning principles. It is difficult to understand its rejection in developing countries where integration is not yet manifest in their space economy.

Meanwhile, neoliberal ideologies prevail as thinking instruments for planning, which initiates paradigm shifts in planning to neoliberal planning paradigm. Neoliberal planning initiates (planning without a plan), in which a market metaphor in development processes determines planning approach. Consequently the relevance of formal planning as a rational instrument to manage the distribution of activities in space diminishes in practice. At the moment pragmatism is the path of growth of planning practice; however, the resilience of formal (form-based) planning is revealed.

Neoliberal planning is real and manifest as investment planning but formal planning still preoccupies the minds of planners. Ironically, both realities are not mutually compatible and a possible synthesis challenges planning scholarship. An alternative option is for planning scholarship to encourage either of the two realities and discourage the scenario where planning initiatives are perceived in part as formal planning instruments but function in reality as neoliberal instruments. This mix-up most likely works against the capacity of planning initiatives to achieve integrated regional development.

However, the choice of interventions in planning for Africa inevitably draws from the need to rework the space economy through integrated regional development. This leaves informal planning out of the options for consideration. For informality in planning to be effectively redressed, ideological change in economic development approach is implied and on the basis of this planning initiatives shall be determined. Driven with the mindset of spatial regional integration, planning initiative should pursue the rational control of the space economy. The use of informal expert knowledge where rational input is impaired is commonplace at the moment. It is found that this practice revolves around favouring funding mechanisms determined by market forces. Thus far current practice has failed to secure integration. It does not support balanced trade-offs with private profitability particularly where sustainable management is the priority.

Given perceived resilience following the spurious evaluation that discredited it, the option of formal planning practice is still open. The resilience of formal master planning is therefore hypothesized.

References

Ayorinde AA, Asamu SO (2000) Integrated physical planning for the maximization of the use of scarce resources: sustainable cities program approach. In: Being paper presented at the 31st annual conference of The Nigerian Institute of Town Planners (NITP) held on 25th–27th October, 2000 at Conference Hall, Shiroro Hotel, Minna, Niger State

Davis M (2004) Planet of slums. New Left Rev 26:1–23

Dawkins CJ (2003) Regional development theory: conceptual foundations, classic works, and recent developments. J Plann Lit 18(2):131–172

Devas N (2001) Does city governance matter for the urban poor? Int Plann Stud 6(4):393–408

Eng TS (1992) Planning principles in pre- and post-independence Singapore. The Town Plann Rev 63(2):163, 183p

Faludi A (2005) Territorial cohesion: an unidentified political objective—introduction to the special issue. Town Plann Rev 76(1):1–13 (In Faludi, A. ed. Territorial cohesion: an unidentified political objective. Special issue)

Friedmann J (2005) Globalization and the emerging culture of planning. Prog Plann 64(3):183–234

Giannakourou G (1996) Towards a European spatial planning policy: theoretical dilemmas and institutional implications. Eur Plann Stud 4(5):608

Harisson P, Todes A (2001) The use of spatial frameworks in regional development in South Africa. Reg Stud 35(1):65

Ilesanmi FA (1998) The role of master plans in land administration and physical development in Adamawa State: the case of greater Yola master plan. In: Ilesanmi FA (ed) Master planning approach to physical development: the Nigerian experience. Paraclete Publishers, Yola, pp 58–65

Isard W (1955) Hybrid land use representation. Available at: http://people.hofstra.edu/geotrans/eng/ch6en/conc6en/hybridlu.html Date of access: 2nd August, 2013

Janin RU (2003) Shaping European spatial planning: how Italy's experience can contribute. The Town Plann Rev 74(1):66, 68p

Janin RU (2005) Cohesion and subsidiary. Towards good territorial governance in Europe. Town Plann Rev 76(1):93–106

Janin RU, Faludi A (2005) The hidden face of European spatial planning: innovations in governance. Eur Plann Stud 13(2):208, 211p

Jiriko KG (2004) Effective management of rapid urbanization in Nigeria: the need for more appropriate planning paradigms. Unpublished PhD Thesis, Department of Urban and Regional Planning, UNEC

Krugman P (1999) The role of geography in development. Int Reg Sci Rev 22(2):142–161

Nadin V, Stead D (2008) European spatial planning systems, social models and learning. disP 172 (1):35, 40–41p

Okeke DC (2005) Towards a global system of planning practice: the Nigerian component. In: Fadare W et al (eds) Globalization, culture and the Nigerian built environment. Ile-Ife: ObafemiAwolowo University, 87p

Onyebueke VU (2011) Place and function of African cities in the Global Urban Network: exploring matters arising. Urban Forum 22:1–21

References

Schoonraad MD (2000) Some reasons why we built unsustainable cities in South Africa. In: Strategies for a sustainable built environment. Department of Town and Regional Planning, University of Pretoria, Pretoria, August 2000

Tawfik RM (2008) NEPAD and African development: towards a new partnership between development actors in Africa. Afr J Int Aff 11(1):55–70

Todes A (2012) Urban growth and strategic spatial planning in Johannesburg, South Africa. Cities 29:158–165

Tunnard C (1951) Creative urbanism. The Town Plann Rev 22(3):234

UNCHS (1996) Sustainable cities: lessons from a global United Nations program. Briefing for the Berlin-Conference on Sustainable Development delivered by Jochen Eigen, Coordinator, SCP, UNCHS

UNCHS (1999) City experiences in improving the urban environment. Nairobi, Kenya

Watson V (2002) The usefulness of normative planning theories in the context of Sub-Saharan Africa. Plann Theor 1(1):27–52

Watson V (2006) Deep difference: diversity, planning and ethics. Plann Theor 5(1):31–50

Watson V (2009) The planned city sweeps the poor away...: urban planning and 21st century urbanization. Prog Plann 72:153–193

Chapter 7
Analysis of Current Reality (Principles and Practice)

Abstract In a comprehensive investigation of paradigm shift in planning, the categories of MCA, perception analysis and SWOT analysis were mobilized to analyse the data generated from planning studies. The functional flow of analysis involved six operations, which investigated compliance to the participatory paradigm of planning. The analyses, which were conducted at national and planning initiative levels, confirmed the incidence of compliance to neo-liberal participatory planning principles but failed to confirm the delivery of integration in planning practice. Regardless of the influence of neo-liberal planning paradigm, there are clear indications of the resilience of formal planning principles. Given the scenario analysis of planning, changes in the meta-theoretical, theoretical and practical context of planning are recommended for planning in Africa.

Keywords MCA · SWOT · Own assessment · Tlokwe · Matlosana · Rustenburg

7.1 Functional Flow of Analyses

The review of the meta-theoretical, theoretical and practical realms of planning, perception and empirical case studies of planning initiatives and country profiles with regard to planning reveals a database that has developed over time and across regional and disciplinary boundaries. The database, which explains the current reality, is subjected to five categories of analyses with intent to determine the status of paradigm shift in planning. These categories of analyses include: using lines of argument to establish the matrix of planning literature, Multiple Criteria Analysis (MCA), SWOT analysis, perception analysis and own assessment using pre-determined template.

The analytical process involved two lines of argument: by timeframe since 1960 and by region covering America, Europe, ASEAN and Africa. The UN-Habitat is included as a major player in the subject matter. The MCA was conducted for the theoretical framework, planning initiative and personal interviews. This set of

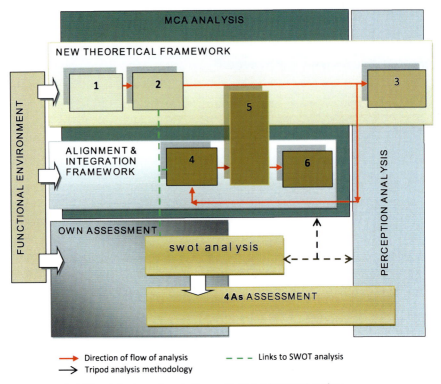

Fig. 7.1 Flow-chart of analysis. *Source* Own construction (2013)

analyses was done to determine compliance to paradigm shift in planning. Two sets of SWOT analyses was done for planning initiatives from select African countries and for case studies of IDPs of local municipalities (Matlosana, Rustenburg and Tlokwe) in South Africa to establish the capacity of IDPs to deliver integration. The same set of IDPs for local municipalities were subjected to perception analyses and own assessment to correlate principles and practice and desired practice). A schematic representation of the functional flow of analyses is illustrated in Fig. 7.1.

7.2 Deductions of Analysis

The deductions of the analyses conducted are hereunder presented. Meanwhile details of the procedure of analyses are contained in Okeke (2015). The deductions are critical in determining the philosophy of planning theory for Africa, at least in the twenty-first century.

7.2.1 Timeframe Analysis

This category of analysis considered the principles and practice of spatial and statutory planning and the informal sector, which is seen as the major challenge of planning in Africa today. Also in the practice section, the evolution of city development in Africa and the incidence of NEPAD were considered. The deductions in this category of analysis were as follows:

 i. The evolution of spatial and statutory planning is cyclical: classic—rational—neo-classic, and symmetrically in practice: pre-modern—modern—post-modern. A similar evolution is noted in urban design: old urbanism—modern urbanism—sustainable urbanism. All three categories of evolutions are influenced by new facts associated with urban growth.
 ii. The informal sector is set to be considered from a spatial perspective following its consideration in the 1970s as a set of activities and in early 2000 as a people-centred phenomenon vis-à-vis the Moonlighting/MML concept,
 iii. The status of spatial planning is unstable. In spite of its systemic isolation and regardless the increasing popularity of non-statutory planning; the resilience of statutory planning is remarkable. The resilience identified is not disconnected from the impact of postmodernism in planning,
 iv. The informal sector remains an economic variable and has increasingly proved to be a morphological factor in city development. Its enhanced popularity in Africa has earned it the impetus to lead the in-formalization of cities,
 v. City development in Africa is in a quagmire. As cities in Africa assume mega-status they increasingly become dysfunctional and progressively lose civic identity,
 vi. NEPAD derives from the neo-liberal tradition of project planning and is currently committed to the RAIDS strategy as operational instrument. The adequacy of this instrument is yet to be determined as the economic growth it is barely able to generate has no relationship with giving form to the urban regions as it is required for the regional integration that is expected from NEPAD initiative.

7.2.2 Regional Analysis

The matrix of sectors examined here are the same with the sectors examined in the timeframe analysis. The regional analysis of planning literature portrays Africa in a peculiar position. As it were Africa is at the receiving end of paradigm changes in planning. Under the strong influence of UN-Habitat activities, the diffusion of paradigm shift from other regions is evident but remarkably statutory planning practice remains resilient in Africa. So planning in Africa is in a kind of dilemma and unfortunately against the backdrop of increasing survivalist informal sector, which does not support accumulation. This scenario draws mostly from the epistemologies of imperialism in Africa. The contrary is the case for other regions where modern informal sector obtains. And planning in these regions, which already have established integrated spatial systems, is not strictly participatory because it accommodates large quantum of urban design (formal planning) practices. However, deductions in this category of analysis were as follows:

i. Europe seems to be more committed to market-oriented planning paradigm than other regions and has gone ahead to articulate ESDP, seemingly in line with participatory planning approach. This scenario is not surprising given the antecedents of Europe in the critique of formal master planning,
ii. A survivalist informal sector that correlates with poverty is commonplace in Africa,
iii. Statutory planning still dominates planning practice in Africa, especially at the local level.
iv. Participatory planning is inversely related to statutory form-based planning and directly related to broad guideline spatial planning,
v. Only the ASEAN region seems not to mingle with city development in Africa. The other regions and UN-Habitat are characteristically involved,
vi. Again, ASEAN countries are seldom associated with NEPAD operations. America and Europe have since positioned themselves to engage especially as development partners. Already, UN-Habitat is practically involved and its supportive roles tend to shift to control as can be argued in its relations with AMCHUD.

7.2.3 MCA Analyses

Approximately four MCAs based on three point value system (positive, moderate and negative) were conducted. The MCA analysis conducted for the theoretical framework of planning (as contained in Table 5.7) generated the following deductions:

i. Multi-dimensional planning perspectives exist in Africa,
ii. Informal planning instruments are used for managing spatial development in Africa,

7.2 Deductions of Analysis

iii. Planning outlooks are diffused in Africa,
iv. Spatial planning is not form-based in Africa,
v. There is nearly a mean distribution of relationship between theoretical and analytical relationship in spatial planning in Africa.

The MCA for planning initiatives analysed four initiatives that were picked from countries at the four cardinal points of Africa. The initiatives include: the City Development (CDS) initiative in Mali (West Africa), the "Shorouk" initiative in Egypt (North Arica), the Opportunity and Obstacles to Development (O&OD) initiative in Tanzania (East Africa) and the Integrated Development Planning (IDP) initiative in South Africa (Southern Africa). The compliance to the new planning perspective, which was analysed, was not even. Relatively, the South African IDP is found to be the most compliant initiative. The deductions in this category of analysis are as follows:

i. The synthesis of data generated indicates that the IDP initiative in South Africa is more compliant with participatory planning paradigm. The initiatives studied cumulatively indicate the dominance of strong compliance although compliance levels are undulating.
ii. Market forces are the dominant determinant factor for land-use management in the initiative studies.
iii. The participatory process is invariably consultative in the initiatives studied, and
iv. Project planning defines the planning framework in the initiatives studied.

Another MCA of theoretical framework sought to determine compliance at national level. A total of ten select African countries were examined (including DRC, Angola, Mali, Senegal, Egypt, Kenya, Nigeria, South Africa, Tanzania, and Ethiopia). The iterative database of selected countries (as contained in Table 3.1) informed the analysis. The performance chart is represented in Graph 7.1. The minimum threshold of performance of each variable in the theoretical framework is anticipated to be 2.4 % because the 42 options (variables) that were assessed generated a frequency of 255 positive appraisals. It turned out that the mean performance of the frequency distribution is 2.5 %. Most of the non-form-based variables scaled the minimum threshold of performance.

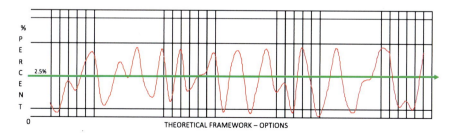

Graph 7.1 Frequency chart of compliance with options in the theoretical framework of spatial and statutory planning in selected African countries. *Source* Own construction (2013)

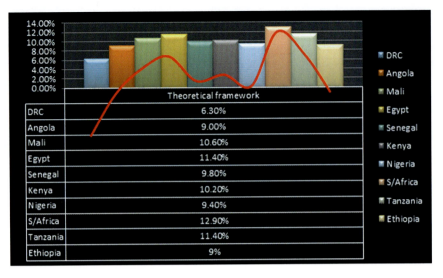

Graph 7.2 Column diagram indicating the level of compliance with the theoretical framework of spatial and statutory planning in selected African countries. *Source* Own construction (2013)

Graph 7.2 represents the descriptive statistics on the horizontal axis. It illustrates the frequency of positive appraisals recorded per country.

The deductions in this category of analysis are as follows:

- The level of mean performance at roughly 2.5 % is above the average threshold of 2.4 % anticipated per variable for a total of 46 variables (options) considered. Thus, the distribution of performance levels is healthy,
- The bulk of issues beneath the mean average line are form-based issues and those above are mainly developmental economics, informal expertise and participatory issues,
- The most compliant country in terms of the new theoretical framework is South Africa and the least is the DRC, and
- The compliance identified is largely in principle, especially in the cases of Tanzania and perhaps Kenya.

It is conclusive that the ten countries are favourably disposed to the new planning theoretical framework for planning, however, in varying degrees with South Africa topping the list. But it was found that all the countries investigated were deficient in the variables that are concerned with form-based planning. Most of the countries are disposed to use informal expertise and participation is anything but interactive. It is noticed that the purpose of planning is focused on economics because planning is increasingly seen as a development activity; hence planning initiatives are evaluated on the basis of plan implementation. The need for regional connectivity and economic integration are some of the other high points of positive disposition towards a new planning theory.

7.2 Deductions of Analysis

The MCA of empirical data from personal interviews on the performance of the IDP initiative in South Africa is conducted based on the appraisal of collated views of respondents and the impressions gathered during the interactive interview sessions. The respondents of the interview include academics, politicians, consultants, and administrators. The deductions in this category of analysis are as follows:

i. Irrespective of compliance with the new theoretical framework there exists a consensus among all categories of respondents that the IDP initiative is potentially a weak planning instrument. In other words, it is not achieving its theoretical role, especially in terms of integration,
ii. The views of academics, consultants, and politicians are symmetrical in collaborating about the weakness identified,
iii. The views of administrators differ from the symmetry shared by the other categories of respondents and tended to be relatively optimistic perhaps due to their positive appraisal of the IDP as a spatial planning methodology with the potential to shape the city and subsequently the space economy,
iv. The four categories of respondents are consistent in their views that the IDP at the moment indicates a negative slant in securing integration and indicates a positive slant as a remote causal factor of 'silo' development given its disposition towards sectoral planning, and
v. Notably the politicians are more incisive (and somehow pessimistic) about the IDP thus signalling their role as potential content drivers in planning.

The analysis of empirical data on the performance of IDP initiative in South Africa derived from planning questionnaire, which followed MCA methodology, indicates positive perception of IDP performance in practice. What this means is that practice is in line with IDP provisions and by implication oriented towards new theoretical frameworks for planning. Figure 7.2 illustrates this staggering compliance, leaving little to speculation.

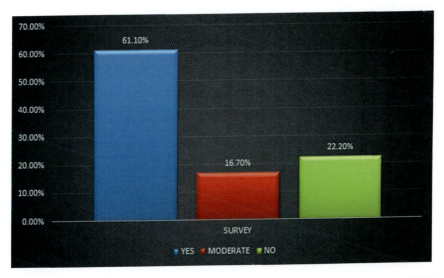

Fig. 7.2 Bar chart of preferred perceptions of IDP performance. *Source* Own construction (2013)

The result of the analysis of planning questionnaire gleaned from the frequency tables indicates:

i. IDP practice is generally perceived to reinforce the new spatial planning theoretical framework,
ii. Administrators and politicians are discretionally more inclined than academics and consultants to perceive positive compliance of IDP practice with spatial planning principles,
iii. Only administrators feel strong in their mutual perception about the positive compliance of IDP practice with spatial planning principles while politicians think otherwise and academics and consultant maintain a moderate position,
iv. Mean average perceptions indicate that only politicians maintain a precarious support for the compliance of IDP practice to spatial planning principles while others, especially academics, are more pessimistic,
v. There is no significant relationship between the perceptions of the different categories as measured by their 'effect size' calculated for planning criteria investigated, and
vi. Overall, the perception of politicians is more sensitive, considering its fluctuation, while that of academics and perhaps consultants is more stable and consistently not positive about the compliance of IDP practice with spatial planning principles. The perception of administrators shares some measure of stability, however, in favour of positive compliance of IDP practice with spatial planning principles.

The result of the personal interview represented is somehow similar to the scenario of perceptions identified in the questionnaire survey. The difference is that in the planning questionnaire optimism about compliance and functionality of the IDP is shared by administrators and not politicians. Perhaps the functionality element is responsible for the twist.

7.2.4 Perception Analyses of Desktop Case Studies

In this category of analysis two control points were held constant. They are the template containing IDP theoretical framework standard and the configuration of desired practice response. The analyses were based on the provisions of IDP documents and educated opinion guided by background information on the IDP initiative. The IDP initiatives that were analysed include the IDPs of Tlowe, Matlosana, and Rustenburg local municipalities.

The analysis of the Tlokwe IDP document indicates that planning is typically event-oriented and engages in extensive consultation. It adopts a holistic multi-sectoral planning approach and seeks legitimacy through legislative provision. It relies heavily on sectoral planning processes. Sectoral analysis was used to elaborate on the existing conditions and asset-based approach was applied in the

7.2 Deductions of Analysis

process to assess available resources. The document stipulated a vision statement which was backed up with medium-term objectives. Priority actions were identified with measurable indicators and subdued spatial integration elements which are typical of participatory planning paradigm. It was found that the Tlokwe IDP document complies very well with the principles of the IDP initiative. The relationship between existing and desired practice is average, requiring adjustments with regards to alignment and integration, its status as a planning methodology, spatio-physical content, and visioning process. The Tlokwe IDP principles comply more fully with desired practice, especially with regards to its appreciation of existing conditions and resource analysis and visionary disposition. Its low points have to do with the definition of the IDP initiative and the inclination towards sectoral planning. The Tlokwe IDP document requires minimal adjustments in other to improve its capacity for integrated planning but the adjustment is critical because it borders on conceptual redefinition. Simultaneously it needs to improve on its status as a planning methodology.

The analysis of the Matlosana IDP initiative indicates that it is properly aligned and integrated with other levels of planning although it inherently lacks integrative attributes for spatial planning. It is committed to participatory processes and identifies with extensive consultation; hence it is sensitive to inclusive planning. Its outlook is more of a development programme than a planning methodology. This reflects in its skewed treatment of existing condition studies which are focused on socio-economic issues without bothering with spatial issues. The disregard for spatial issues is further reinforced with the status of Spatial Development Framework (SDF). The SDF is not empowered to determine land use; in fact it is not available because attention is focused on land reform projects. The compliance of IDP practice with its principles is not impressive. A 48 % compliance level is identified. The bulk of criteria for compliance are indeterminate. It was found that the relationship between existing and desired practice fell, relative to the relationship between principles and practice. Lots of weaknesses are identified in the relationship especially as it relates to their disposition on process, nature of planning, definition of planning as methodology, focus on the IEP, attention on cross-cutting issues, duration of planning objectives, and the status of the SDF, etc. The relationship between principles and desired practice is much better. Areas of weakness are drastically reduced; however, adjustments are required in the planning process, the nature of planning, the planning outlook, definitions of planning, cross-cutting issues, the institutional plan and plan-approval processes. Overall there are some critical observations on the Matlosana IDP initiative. The pre-2004 SDF had no link with higher order plans. The SDF is usurped through IEP, land-use schemes which tend to duplicate the effort of the SDF, financial plans, institutional plans and sectoral plans. In the circumstances, the IDP initiative tends to regard integration as a summation of priority projects identified on ward bases. There is a very strong need to reposition the SDF and enhance the possibility of integrated planning and development in the municipality.

The Rustenburg IDP initiative aligns properly with higher order plans. It adopts activity-centred processes and engages in extensive consultation to deliver sectoral planning. Its outlook is implementation-oriented due mainly to its link with the budget. The Rustenburg IDP seeks legitimacy through legislation and it identifies with the definition of the IDP that emphasizes a planning methodology. Therefore, it is visionary given its long-term objectives. Remarkably it is linked with the CDS initiative. Its existing condition studies are good but its resource analysis is poor. Cross-cutting issues attract lots of attention, a scenario that is perhaps responsible for the identification with sector plans. There is no indication of an integrated SDF, although there are indications that its SDF shall serve as a guide to decision-makers. Spatial integration is potentially endangered. It was found that the Rustenburg IDP practice properly interprets IDP principles. The positive trend towards compliance indicates the need for upward adjustment of current indeterminate positions classified as moderate. Positive adjustment means the reduction of negative relationships. The anticipated adjustment is almost the situation represented in the relationship between practice and desired practice. Similar adjustments are required for the relationship between principles and desired practice. Upgrading of weak relationships by at least ten percent is critical to achieve positive trends towards positive relationships.

Noteworthy in the analyses of the IDP initiatives is the consistently high percentage for positive relationships in all three scenarios examined. The percentages of positive relationships are higher than those of moderate relationships and those of moderate relationships are in turn higher than those of negative relationships. What this implied is that positive connections or linkages exist between principles, practice and desired practice; although to improve the linkage there is room for adjustments. The adjustments are prone to draw upon aspects of the IDP that exhibit weak relationships and these aspects are mainly linked with spatial integration issues such as the status of the SDF, integrated rather than sectoral planning, etc.

The deductions in this category of analysis are as follows:

i. The performance of IDP frameworks vary but they all maintain positive compliance with the new theoretical framework,
ii. The relationship between existing and desired practice is positive although it varies in relation with compliance with principles,
iii. A strong positive relationship exists across the board between principles and desired practice,
iv. Weak relations are at a low ebb and more pronounced in the relationship between practice and desired practice,
v. The capacity of IDPs to achieve spatial regional integration is generally below average although a comparative advantage exists between IDPs.

7.2.5 SWOT Analyses/Own Assessment

SWOT analysis was initially done for the planning initiatives in select African countries. The SWOT analysis was targeted to measure the potentials of the initiatives to support spatial integration. The deductions of the SWOT analysis are as follows:

7.2 Deductions of Analysis

i. The IDP initiative exhibits the best disposition towards compliance with the new spatial planning theoretical framework that has developed since the 1980s.
ii. Planning initiatives in selected African countries are generally weak spatial planning instruments. They are all weak in spatial integration but somehow strong in resource mobilization.
iii. Planning initiatives in selected African countries cumulatively indicate very low percentages of a strong capacity to deliver spatial regional integration in the continent, and
iv. The compliance of planning initiative with participatory planning paradigm is on a higher level in South Africa and Egypt compared with the other initiatives that were studied.

Overall, the new theoretical framework which has developed since 1960 is found to influence the conception of planning initiatives in Africa. Among the four African planning initiatives examined, the IDP in South Africa seems to be more compliant with the new theoretical framework. Although the new planning participatory planning perspective seems to be gaining some foothold, it is not delivering integration.

The second SWOT analysis was done for IDP initiatives in South Africa with intent to determine the capacity of IDP initiatives in South Africa to deliver integrated spatial development. The IDP culture has more or less influenced the content of planning rather than the modelling of planning methodology. The IDP methodology, which has remained within the confines of traditional budgetary planning approach, is riddled with bureaucratic red tape. The deductions of the SWOT analysis are as follows:

i. The IDP is realistically an investment planning instrument,
ii. The IDP as an instrument for spatial planning is contestable,
iii. Not all IDP documents are visionary,
iv. The IDP conceptually lacks integrative elements. There is a need to mainstream integration,
v. The SDF is a misplaced strategy because the SDF is conceptually invalid under an IDP culture,
vi. The IDP lacks authority in practice.

At this point, the prospects of the IDP to provide integrated planning possibilities are growing dimmer. Own assessment reconsidered the IDP instruments to establish their capacity to achieve integration. To this end, lots of syntheses of already established relationships were engaged. The deductions of own assessment, which is evidence based, are as follows:

- The pattern and quality of IDP documents vary.
- IDP planning is potentially weak and vulnerable as a spatial planning instrument.
- IDP documents are fairly well related to the IDP theoretical framework.
- The IDP is more advanced in theory than in practice.

- Patterns of relationships between principles, practice and desired practice are not identical but fairly similar within and across the municipalities.
- Levels of relationships between principles, practice and desired practice within and across municipalities are consistently positive.

In summary, the IDP is more of a management instrument than a planning instrument. It is supposed to provide integrated planning but unfortunately the various sectors in the planning machinery work in isolation from each other, each protecting its mandate. Hence, reactive planning is encouraged and the integrative element is lacking. The planning outlook of IDP is existential and incremental. It is not necessarily proactive and futuristic as integrated planning provides.

7.3 Test of Hypothesis

Qualitative and quantitative research was used to test the hypothesis.

Following the inception of neo-liberal theory, the literature review established the theoretical framework for planning since the 1980s. The new theoretical framework emphasized process-planning which compromised formal planning rationality and tended to promote informal planning (c.f. 6.5.2). The compliance of country profiles in Africa with the theoretical framework was analysed. The result established compliance between the analytical framework of planning and the new theoretical framework in principle and not necessarily in practice (c.f. 6.7). South Africa topped the list of ten countries sampled (c.f. 6.7). The compliance perceived in the African country context was further examined by relating the theoretical framework to planning initiatives in selected African countries (c.f. 6.7). Compliance was further established, however at varying degrees but generally above average (c.f. 6.7). The South African IDP initiative was found to be more compliant (c.f. 6.7). Form-based elements performed below average. In the very unfavourable circumstance it maintained a precarious presence thus signalling resilience (c.f. 6.7).

On the other hand, the matrix of literature by timeframe illuminates the resilience of statutory planning underpinned in cyclical trends identified in the evolution of planning principles and practice. The resilience connects with the effect of post-modernism in planning. The status of planning is identified as being relatively unstable. The matrix of literature by region indicates the dominance of statutory planning in the planning practice in Africa contrary to the scenario in Europe. The literature analysis also identified the peculiar relationship between participatory planning, statutory planning and broad guideline statutory planning. The inverse relationships identified indicate formal planning as a resilient player in the neo-liberal environment of participatory planning. Furthermore, the review of secondary data on country profile indicates that the side-lining of formal planning is more or less active in principle and not in practice. Overall literature indicates

7.3 Test of Hypothesis

multi-dimensional perception of planning perspective shared between participatory and master planning perspectives. Most participatory perspectives are intended for project planning, especially for renewal projects. In effect the perception of planning perspectives in Africa is diffuse, thus signalling the resilience of a form-based (statutory) outlook which is consciously being subordinated.

For purposes of quantitative research, the IDP initiative was subjected to an empirical investigation in the context of planning in South Africa. Compliance with neo-liberal attributes was confirmed; however, not without attendant issues that border on functionality of planning initiatives as spatial integration instrument and the status of planning rationality. The perception of the IDP as a spatial planning instrument varied, with administrators maintaining consistent positive appraisals and remarkably followed by politicians. Mixed perception suggests resilience of issues outside the mainstream of neo-liberal planning. This is confirmed in the impressions garnered from the personal interviews in which the administrators were favourably disposed to recognizing the spatial planning principles that drive IDP practice.

Further analysis to determine the resilience noticed was conducted with the database on the perception of IDP practice. The database drew from a total of thirty-six questions that examined formal (form-based) and pragmatic (non-form-based) planning themes. The questions were evenly spread for probing the two categories of planning paradigm and those for formal (form-based) planning are isolated (see Appendix A.1). The analytical approach sought to determine the performance of each category of planning in the percentages of holistic perception identified earlier. Consideration in this analysis is given primarily to the ratio of percentages of the positive (that is "Yes") perceptions. Ratios above zero indicate resilience.

The calculated ratio of **1.4** deduced from the split percentages of "Yes" perceptions (see Appendix A.2) confirms the existence of a very strong resilience of formal (form-based) planning attributes in the perceptions of analytical frameworks of IDP planning initiatives. On the other hand the distribution of the disposition of the four categories of respondents derived from the database (see Appendix A.3–A.5) is determined from the matrix of calculated ratios (see Appendix A.6). It is deduced from the ratios that the perceptions of administrators, politicians and academics clearly and consistently identify with the existence of a very strong resilience of formal (form-based) planning attributes in the analytical framework of the IDP planning initiative. The perception of consultants, judging from calculated ratios is confirmed as being strong and almost very strong. Overall, the resilience of formal (form-based) attributes in IDP planning initiative is roundly confirmed. However, this fact pales into insignificance and fails to make an impression given the overwhelming popularity of neo-liberal participatory attributes of IDP initiatives.

Therefore, from the foregoing it can be deduced that the research disproves the null hypothesis. Thus, in the perception of planning initiatives within spatial systems in Africa, form-based planning attributes are significantly resilient. What this implies is that form-based planning is a character trait that identifies spatial planning. As participatory (neo-liberal) planning attempts to do, it is an exercise in futility to ignore it. The sooner this fact of planning is recognized and reconsidered in the conception of planning theory, the sooner spatial planning will be back on track to achieve spatial integration.

7.4 Scenario Analysis

The scenario analysis contemplates the impact of changes in the meta-theoretical, theoretical and practical levels of planning on the basic scenario of planning. The basic scenario is framed on neo-liberalism as the meta-theory of planning. Neo-liberalism advocates for market force as determinant of planning based on its general ethical precept of private profitability. The neo-liberal planning theory occupies the theoretical level. It advocates a peculiar participatory planning perspective that is diametrically opposed to the classic participatory planning paradigm. Its objective is not necessarily in securing participation but in decentralising planning decision ruled by market force principles. At the practical level planning initiatives assume the outlook of investment planning thus redirecting attention from the city to the people. Given this structure, the basic scenario of planning practice (c.f. 3.4) tends to emphasize environmental management rather than urban planning. The direction of planning attention away from the spatial systems as the subject of spatial planning work to exacerbate the urban crisis particularly urban sprawl and its attendant problems. Thus, spatial integration lost impetus as the object of spatial planning.

The basic scenario of planning is built on the rapid urbanization argument, which is called to question. What if there is no rapid urbanization in Africa as Potts (2012) postulates? In the same vein, it relies on the infrastructure hypothesis to prospect for productivity and the development of urban economy. The infrastructure hypothesis is an economic model and not a planning recipe. What if productivity is based on form and function principles of planning? The affirmation of any of these questions can potentially shift the status quo of planning. Meanwhile, regardless of the provision of the precondition of democratic political governance, the basic scenario is rated below standard compared with performance in other regions of the world. Urban productivity decline is manifest and it is established that African cities occupy a very low place and function in the global economy (Onyebueke 2011).

The scenario analysis of planning holds democracy and good governance in Africa constant. This position is held with some reservations because of three trends associated with evaluating democratic governance in Africa. First, the political institutions are consistently adjudged to be weak, thus requiring capacity-building; second, increasing association of the political system with participatory process; and third, it seems the more responsive African governments are to imperial control the more they are adjudged to be democratic in the perception of the global north. These attributes are bridges in the global political economy that secure the incapacitation of the bargaining position of Africa. Also held constant is the

7.4 Scenario Analysis

consideration of African renaissance as the object of planning. African renaissance is held as a theory of development that defines the mission of cities in Africa. The African renaissance elements as well as the knowledge of the resilience of formal planning are held as new facts in the reconsideration of the status quo in planning.

There are six combinations of possible changes in the meta-theoretical, theoretical, and practical context of planning. The first combination considers a change in the meta-theory of planning while the theoretical and practical context remains constant. A lot depends on the direction of change of the meta-theory but in so far as the planning perspective remains constant the change will not generate much difference in the status quo of the development of spatial systems in Africa. The second combination where the meta-theoretical and theoretical changes and the practical context remain constant, moderate changes are anticipated in the development of the spatial systems. In this scenario, it is not likely the practical context will be resilient. The third combination where there are changes across board, automatic change in the development of the spatial system will occur. Where the change in the meta-theory correlates properly with ethos of African renaissance particularly nation building, the change is likely to be positive.

The fourth combination where there is no change in the meta-theoretical and there are changes in the theoretical and practical context, the status quo in the development of the spatial systems may rework significantly but most likely in favour of imperial development. The fifth combination where there no changes in the meta-theoretical and theoretical and the practical context changes, isolated and insignificant changes may occur in the development of the spatial systems. The sixth combination where there are no changes across board, the status quo in the development of the spatial system remains.

The first, second, fifth and sixth combinations where the practical context does not change are the most likely to be resilient because the promise to sustain the status quo. The second, third, fourth and fifth combinations where the theoretical changes, the prospect of changes in the status quo is feasible especially where the practical context is also disposed to change. But the change anticipated is likely to be within imperial development. Epistemologies in the development of spatial systems cannot be redressed in these scenarios. The prospect of change is highest in combinations where the meta-theory of planning changes and it is optimum where there are changes across board.

The incidence of new facts in environmental determinism will support changes in the practical context. This works best where change in the meta-theory is also anticipated. It is noteworthy that the meta-theory influences attitude towards the use of environment either around the philosophy of eco-centric or techno-centric mode of modern environmentalism. Given the basic scenario, any change in the meta-theory of planning will tend towards eco-centrism, which is collaborated in the principles of African renaissance. Thus, the modelling of the development of spatial environment is inevitable and this enhances the stakes of the resilience of formal planning. In the circumstance the option of changes across board is recommended for positive change in the basic scenario of planning in Africa. The feasibility of this option demands changes in trends identified in the evaluation of democracy and good governance in Africa. In other words, the anticipated change of status quo is in the first instance a political mission then a theoretical redefinition and ultimately a practical reality.

7.5 Conclusion

The compendium of analysis so far indicates compliance with participatory planning paradigm and the difficulty of planning initiatives to deliver integration. The need to deliver integration tends to draw attention to the resilience of the formal planning paradigm. the analysis of the database from the empirical study concludes the analytical process meant to verify trends in participatory planning. In this, the result of personal interview confirms the findings of the questionnaire survey, especially as it relates to the correlation of perceptions among the various categories of respondents. It is generally perceived that the practice of the IDP initiative does not indicate the same level of sophistication as its theoretical background. The appreciation of the IDP concept was found to vary depending on capacity and this reflects on the quality of initiatives. No correlation was found in the perception of the different categories of respondents in the questionnaire survey and this further highlights the enigma of the concept. But the general impression is that the IDP exhibits a weak capacity to deliver integration. Remarkably, respondents of the personal interview confirmed that integration was not being achieved. Across the six levels of analysis that was conducted, integration was at stake. It was difficult to find an interface between integration and process-planning initiatives. The challenge ahead is to find this interface—that is, if it is at all feasible. It is most likely that solace could be found in synthesizing equilibrium between political analysis and technical analysis in planning. This is found to be the practical problem with the IDP initiative in South Africa—apart from enhancing form-based planning orientation in reconsidering the initiative.

A summary of the findings of analyses of planning perspectives and initiatives are contained in Appendix A.7. Thereof the research conclusively postulates that planning initiatives that are compliant with participatory planning paradigm do not have the capacity to deliver integration in development processes. It further postulates disconnect between participatory planning theory and spatial modelling of space economy. This is as a result of the relegation of formal planning principles. Regardless of the hegemonic influence of neo-liberal participatory planning paradigm, there are clear indications of the resilience of formal planning principles and practice. This knowledge base cannot be ignored because it synchronizes with the territorial planning outlook identified for the delivery of African renaissance. Therefore, an alternative planning theory that leans towards strong intervention ruled by the principles of equity and sustainable development is high in the agenda of planning for Africa.

References

Okeke DC (2015) An analysis of spatial development paradigm for enhancing regional integration within national and its supporting spatial systems in Africa. Available at: http://dspace.nwu.ac.za/bitstream/handle/10394/15485/Okeke_DC.pdf. Date of access 19th June 2016

Onyebueke VU (2011) Place and function of African cities in the global urban network: exploring matters arising. Urban Forum 22:1–21

Potts D (2012) Whatever happened to Africa's rapid urbanization. Africa Research Institute. Available at: www.researchgate.net/.../254456370 Date of access: 9 Nov 2015

Part III
Synthesis and Statement of New Theory

Chapter 8
African Renaissance

The Subject of Planning (in Africa)

Abstract In the African context where political realities make equity a critical variable for new economic and spatial entities, the theory of African renaissance provides the object for planning and the mission for cities. African renaissance is up against a new agenda of sharing African market for Euro-American goods and services. Hence this theory seeks the primary objective of integrated regional development, which connects with sourcing enhanced productivity through the introversion of the economy of urban Africa. It demands the restructuring of the urban form through territorial planning and in so doing hopes to generate communities of African renaissance of the twenty-first century with capacity to surmount contemporary urban crisis in Africa. Hitherto communities of African renaissance of the Middle Ages, with diverse systems of behaviour and belief bound with spiritual values, who ousted the backward Bushmen and Pygmies, overcame similar obstacles and founded cities and built states and empires.

Keywords Africa · Renaissance · Euro-American · Productivity · Dependency · Colonialism

8.1 African Renaissance—a Theory of Development

The pedigree of African renaissance as a concept of blossoming Africa dates back to the tenth century when Empire and Kingdom building, particularly Sudanese Empires, pioneered tropical civilization. The African communities, with diverse systems of behaviour and belief bound with spiritual values, who ousted the backward Bushmen and Pygmies, overcame obstacles posed by harsh environmental setting in the sub-region and founded cities and built states and empires. Prior to this period, memories of Greek colonialism in North Africa dating from the fourth century witnessed the emergence of flourishing ports, such as Carthage and Alexandria.

The modernization theory, the movement of 1950–1960s rooted in capitalism, intercepted the African renaissance of earlier epoch. Modernization, which ensued, impoverished Africa through colonialism and imperialism by the West and this trend is with us today as the East takes its turn to deplete the continent's resources such as oil and minerals (Matunhu 2011: 67). Matunhu further states that the 'ideas of modernization impoverished Africa. The theory failed to recognize the creativity and initiative of Africans'. Discontentment with the modernization theory in the 1950s precipitated new strands of thinking which resulted in the dependency theory. The dependency theory positioned Africa to specialize in marketing raw material while the developed world market finished products. The theory engineered Africa's poverty position and ultimately, according to Matunhu, 'Africa lost its right to determine its way to development'. Indeed social anthropologists consider the dependency theory to be both pessimistic and structural (Matunhu 2011: 68).

The antithesis to the modernization and the dependency paradigms is the emerging African renaissance theory. The theory is founded on African values and norms that are the very building blocks of African life (Matunhu 2011: 71). Therefore, its theoretical framework is based on African value system and African identify. The earliest formulation of this theory is originally linked with Africans in the Diaspora in response to mistreatment of the *Negro* in the Americas and in other places where Africans were carried off to slavery. Within the African context, the rationale for the theory drew from the lingering presence of neocolonization and imperialism, which is largely responsible for economic stagnation in Africa. The struggle against imperialism in Africa was a struggle for African independence and to that extent for an African renaissance (Nabudere 2001). This mindset and viewpoint is rooted in Pan-Africanism, a concept for a united Africa and for the liberation of its peoples from colonialism (Nmehielle 2003: 426). The analyses of the theory increasingly relate people and activities to industrial and economic development thus expressing skewed emphasis on economic integration for growth.

The emergence of the modern concept of African renaissance coincides with the decade of African independence in the 1960s. Renewed attention on the concept was born following the progressive regaining of power by the indigenous people in the nations of Africa (Creff 2004: 3). The concept was said to be first articulated by African Historian and Anthropologist, Cheikh Anta Diop in a series of Essays beginning in 1946 which are collected in his book—Towards the African Renaissance: Essays in culture and Development-1946–1960 (Ndu 2014). Based on Diop's submission African Renaissance is the concept that African people and nations shall overcome the current challenges confronting the continent and achieve cultural, scientific and economic renewal. Subsequent submissions polarized into two contexts defined by those developed outside and within South Africa. The scenario within South Africa is more profuse and dominated by the perceptions of Thabo Mbeki.

Although it is claimed that the African renaissance concept has been subjected to debate amongst African academia, contributions that are divorced from the views of Mbeki are limited. Van Niekerk (1999) made one of such rare contributions and he posits that the African Renaissance is a concept borrowed from the European Renaissance and refers to 'the period of cultural and intellectual achievement that followed the era of late scholasticism'. According to Creff (2004) Van Niekerk (1999) further 'pointed out that by comparison, the European Renaissance developed over a period of at least 150 years, a time frame that is unrealistic given the severity of Africa's problems. Additionally, the European Renaissance had 'small beginnings' and 'depended on the work of a very small minority', whereas Africa needs a widespread impact on its difficulties for the achievement of significant changes. Some African scholars therefore argue that it is a borrowed concept from experiences unique to Europe thus rendering it irrelevant to Africa.

Cossa (2009) attempted a conceptual analysis of African renaissance with intent to legitimize the use of the term 'African renaissance'. Cossa established that there are normative values and peculiar traditional civilization that are attached to the concept of Africa. These unique values provide legitimacy for the use of the term 'African'. The term 'Africa' according to Cossa (2009) describes those individuals, and things associated to them, who are native to Africa and can (in one way or another, but not necessarily in terms of genealogy), trace their ancestry to indigenous African people-groups. Thus Africans are entitled to use African to describe, exclusively, their indigenous experiences and past traditions and civilizations just like their European counterparts do. Examining the term 'renaissance' presented more difficulties. The African Renaissance presents similar trends to those of the European Renaissance. Many African leaders and scholars use the term 'African Renaissance' limiting it to reclaiming the validity, in a global sphere, of indigenous forms of African civilization and languages. This draws attention to the regeneration of Africa that reflects African innate creativity, rather than creativity acquired from Europe or America. This puts the application of the term 'renaissance' in a fix.

Renaissance is a French term for 'rebirth'; thus, to use such a term to mean only that which it means, i.e. 'rebirth', is fair. However, Cossa (2009) observes that in light of the arguments made by the very movement, theory, philosophy or era (whatever the perspective may be on the nature of the African Renaissance) that Africans are attempting to describe, it is not legitimate to use a European term for an African phenomenon. The concept needs to be reevaluated, particularly in academic discourse where the opinions of the intelligentsia are stamped and propagated, and a term must be found that characterizes the phenomena in an African way (Cossa 2009). This calls for African term such as 'Ubuntu' to be used to describe African phenomena.

For contributions that are developed outside the South African domain, the validity of the concept of African renaissance to address African regeneration is called to question although the arguments in favour of this position seem to dwell largely in making semantic conceptual link between African renaissance and the African-nation besides the concern for securing innate African creativity. Otherwise the concept of 'African renaissance' is held by some critics to be discomfited given the foreign origin of the term 'renaissance' and its imperial connotations. However, for the leaders of the African Union and Heads of African governments, African renaissance is a perfect phrase for sloganeering but for the black people of Africa, including its Diaspora, it is the road to salvation from imperialism, want, poverty and disease (Ndu 2014).

According to Van Kessel (2001: 44) many South African adepts of the Renaissance seem remarkably unaware of the long pedigree of the African

Renaissance concept. The concept of African renaissance in South Africa domain is built around the contributions of Thabo Mbeki in late 1990s. Most of the African renaissance debates took place in black intellectual circles. Incidentally the debates are critical of Mbeki's version of conceptualizing African renaissance theory. In all of the happenings Van Kessel (2001: 44) discerned three interpretations of the African Renaissance in South Africa: as an agenda for modernization, an agenda for neo-traditionalism, and an agenda for Africanisation. In all three cases, the African Renaissance is used to fill an ideological vacuum in the post-Cold War world and according to Boloka (1999) it is linked to Thabo Mbeki, 'who articulated it as a means to Africa's empowerment'. Nabudere (2001) argued that the concept is a useful tool in the struggle of the African people to redefine a new political and ideological agenda of pan-Africanism in the age of globalization.

Mbeki's vision of Africa's rebirth was initially launched as a pan-African vision, which is the popular position in black intellectual circles. This vision was inspired by a largely mythological interpretation of African history (Van Kessel 2001: 46). The vision is driven by 'the rediscovery of our soul, captured and made permanently available in the great works of creativity represented by the pyramids and sphinxes of Egypt, the stone buildings of Axum and ruins of Carthage and Zimbabwe, the rock paintings of the San, the Benin bronzes and the African masks, the carvings of the Makonde and the stone sculptures of the Shona'. (Mbeki cited in Van Kessel 2001: 46).

At the turn of the century Mbeki's renaissance vision truncated and focused within South Africa. The new outlook of Mbeki's vision of the African Renaissance includes joining the information superhighway, the emancipation of women, debt cancellation, improved access to international market for African products and sustainable development (Van Kessel 2001: 45). Although it did not lose its pan-African ambition it became a secular modernizing program that is political and economic in focus and that pays comparatively little attention to cultural dimensions. Van Kessel (2001: 47) summarized Mbeki's position in the following words:

> Mbeki's vision differs in at least two important aspects from previous Renaissance philosophies, developed by previous generations of African intellectuals over the past century. His focus is on accelerated economic development, on joining the global economy, not on culture or the spirituality of the African soul as was common in previous Renaissance waves. Secondly, unlike his predecessors, Mbeki does not invoke a unique African genius.

Mbeki is not interested in notions of African uniqueness but in obtaining Africa's full share of universal progress. Thus he shared Universalist views which are neither anti-capitalist nor anti-globalization per se (Van Kessel 2001: 47). His views were not universally shared and this is not unconnected with the parallax earlier identified in the perception of the term 'renaissance'. His modernizing views do not blend with the culturalist interpretation used to inform the agenda of traditionalists and neo-traditionalists as well as Africanists. Given the culturalist interpretation, African Renaissance as a legitimization for neo-traditionalism becomes an exclusivist notion and as an agenda for Africanisation is often understood as a 'blacks only' thing. As an agenda for modernization traditionalist still invoke Africa's heritage to regain power and privileges although this often times spell bad news for gender equality entrenched in Mbeki's modernizing program.

Reactions towards African renaissance vary on the basis of which three schools of thought are discerned. The schools of thought vary between Afro-optimism and Afro-pessimism with Afro-realism taking the middle ground (Botha and Pierre 2000). The first school of thought seems to base its views on the renewal of Africa predominantly on normative considerations. The overwhelming structural problems in Africa colored the outlook of the second school of thought thus compelling it not to share the optimism of the first school. In the context of global trends the Afro-pessimism school literally calls African renaissance to question. This school is dominated by the political left, critical academics and radicals who view African renaissance simply as a pseudonym for American and European imperial ambitions into the Eye of Africa (Terreblanche 2012: 68). For them the christening of the concept is critical and regardless the renaissance philosophy Africa is still being subjugated to the so called North bound Gaze resulting in the denial of Africa choosing its own experiences, identity and interpreting its own history and politics (Oelofse 2013). The Afro-realism school can be said to perceive African renaissance as a third moment in Africa's post-colonial experience; the first is decolonization and the second is the 1990s neoliberal insurgence. For this school and perhaps the Afro-optimism school, which is associated with Mbeki and his ANC government, African renaissance is part of a broader anti-imperialist movement. But 'to the government and businesses of South Africa today, African Renaissance is a brand name, that can be effectively deployed to promote capitalism and neo-imperialism, but to the numerous black miners and people of South Africa, African renaissance is a clarion call for Black emancipation' (Ndu 2014).

Regardless the schools of thought, the perspectives of African renaissance differ. Again three perspectives are identified: globalist, Pan-African and Culturalist perspectives. The globalist perspective emphasizes political and economic renewal. This is associated with Afro-realism school and perhaps the Afro-optimism school. The pan-African perspective links African renaissance to Pan-Africanism, yet seen as 'an attempt to mobilize Africans to unite against the tyranny of colonialism by redefining an African identity and freedom independent of colonial influence' (Cossa 2009: 5). The tendency of the Afro-pessimism school to identify satirically with this perspective is high although the *African Renaissance* is a reigniting of the spirit of Pan-Africanism. The debate within this circle is centred on the question: Is the renaissance a Pax Pretoriana thinly disguised as a Pax Africana? The culturalist perspective earlier mentioned informed, as it were, by ethno-philosophy, sees the African renaissance as a movement for a return to the 'roots' (Maloka 2001). This perspective encourages the exclusivist notion of African renaissance. The Afro-optimism school identifies with this perspective. Ultimately the exclusivist and Universalist notions of African renaissance exist in South Africa, both contesting for relevance with incisive arguments that are mindful of issues of identity, content, and neo-imperialism. These events are based on the periodicisation of the renaissance discourse from 1994. Summarily, as Nabudere (2001) noted, the usage of the term 'African renaissance' in South Africa is 'Janus-headed' meaning:

> On the one hand, it reflects the mainstream political elite concern in South Africa for an African national identity against the background of an alienating apartheid system, which tried to depict South Africa as a white man's country. From that standpoint, South Africa was not part of the African continent socially, politically and culturally. On the other hand, it can also be seen to express the ANC's concern to win corporate capital in South Africa to its programs in the age of globalization. It can be said that the struggle between these two understandings of the African renaissance will determine the direction in which the movement will proceed (Nabudere 2001).

In the circumstance, attaching imperial content to the meaning of the term 'Renaissance', which introduced undue consideration of syntax and controversy in the renaissance discourse, is somehow diversionary. Renaissance can be applied literarily to mean (as formulated in the English Language Thesaurus tool available on Microsoft Word): rebirth, new start, new beginning, resurgence, revitalization, revival, regeneration, recovery, and reawakening. The 'rebirth' option applies most suitably to address what is considered the central element of renaissance, which is economic growth free from imperialism and neoliberal principles, regardless contextual forces—global or regional hegemonies. Although, according to Mbeki (as cited in Makgoba 1999), 'we cannot win the struggle for Africa's development outside the context and framework of the world economy'. Nevertheless, growth as it applies here implies two categories of integration—economic and spatial integration. Note however that both categories of integration can either be formal or informal. The formal option, subsumed in creative pan-African urbanism, is emphasized for African renaissance.

The orientation of Afro-realism school best captures the Pan-African perspective of renaissance outlined here. It is against this backdrop that African renaissance is defined. Hitherto a couple of attempts have been made to define African renaissance from institutionalist and independent sources. The African Renaissance Institute (2000) defines African Renaissance as

> A shift in the consciousness of the individual to reestablish our diverse traditional African values, so as to embrace the individual's responsibility to the community and the fact that he or she, in community with others, together are in charge of their own destiny.

This definition relates to the change of mindset of the individual, specifically freedom from colonial mentality. The independent definitions are built around Mbeki's understanding of African renaissance. In the words of Botha and Du (2000):

> The latest understanding of the African Renaissance, or rather- the African Renaissance as President Mbeki understood (and desired) it to be, relies heavily on neoliberal economic principles of the Free Market System and privatization, systems that have been in place since the Apartheid era; sustaining the vicious cycle of poverty and inequality (Terreblanche 2012). This statement is also supported by the fact that South Africa's extremely wealthy elite are exceedingly welcoming of the idea of an African Renaissance (Vale and Maseko 1998: 279).

Mbeki's drift as outlined in the above passage over-stepped into accepting neoliberal principles. His idea is therefore indifferent to globalization tendencies, which involves continued exploitation of African's resources. The perception of South Africa as a regional hegemony could have informed this position and more so in the context of the willingness to enlist South African's resources to renew the African world (Muchie 2004: 148) and secure leadership status for South Africa in the African region. Mbeki posited that leadership development is necessary to ensure the advancement of the African Renaissance (Creff 2004: 3).

Literature indicates that African Renaissance is not a policy per se but an idea or ideology that gives way to various policies (Ramose 2007). The aims of African renaissance are contained in its policy framework. According to Mbeki (as cited in Makgoba 1999), 'one of the central aims of the African Renaissance is the provision of a better life for these masses of the people whom we say must enjoy and exercise the right to determine their future. That renaissance must therefore address the critical question of sustainable development which impacts positively on the standard of living and the quality of life of the masses of our people'. This reflects a globalist view. On the other hand a Pan-Africanist view also associated with Mbeki envisages 'an African continent in which people participate in systems of governance in which they are truly able to determine their destiny and put behind the notions of democracy and human rights as particularly Western concepts' (Nmehielle 2003: 427). The former dominates the policy framework of African renaissance, which is contained in NEPAD initiative.

The NEPAD initiative has been severally criticized strictly on account of its globalist perspective of African renaissance. In the words of Nabudere (2003), 'the African people must continue their struggles for the dismantling of the old economic order instead of compromising with it if we are to address the kind of issues that NEPAD tries to address and which it avoids to address'. Unfortunately some of the stated and meaningful objectives of NEPAD initiative such as to halt the marginalization of Africa in globalization process are tactically being supplanted with Millennium Development Goals (MDGs). This has the effect of influencing the framework of spatial information system for African renaissance. The core dataset anticipated for the delivery of African renaissance includes socio-economic information, service need and provision, and development funding. These dataset are carefully chosen to facilitate the definition of priority areas for foreign aid.

The existing frameworks are focused on objectives that have been set for Africa through the signing of protocols and the implementing of specific strategies so that the necessary spatial information for decision-making can be identified (Schwabe 2001). The frameworks also function to develop an integrated approach to planning and analysis (UNEP 1999). Already there exists, for example, the Africa Information Society Initiative (AISI), an action framework to build information and communication infrastructure in Africa while the United Nation's Development Assistance Framework (UNDAF) is to coordinate development funds—perfect setup to deliver imperialism. The impression is given in Creff (2004) interpretation of Mbeki's views that:

The achievement of the goals of the African Renaissance are not possible without the support of First World nations in providing aid for immediate needs, but more importantly, in assisting in the development of African leadership to contemporary effectiveness and the ability to reproduce leadership across the continent that will assist followers to harness the potential of the continent.

There are matters arising from African renaissance discourse in South Africa. In the first instance, the controversy in the definition of an African is a South African issue. The possession of citizenship of any African country qualifies any individual as an African irrespective of color or race. It does not matter if such individual share the philosophy of Pan-Africanism or not. Pan-Africanism is a global perspective of local development ideologies in the African domain. The attempt to present 'ubuntu' as central element of African renaissance is noted but recognized as a localized effort. Again the 'ubuntu' craze is a regional sentiment shared in Southern Africa. Each region in Africa has its own 'ubuntu' in different moulds. Also the government and people of South Africa are propelled by different forces in their consideration of African renaissance. While the perception of government is driven by growth visions, which is not properly defined, the people are still reactive to the scares of Apartheid. Furthermore the conceptualization of African renaissance is driven by economic integration in total disregard of spatial integration. The element of territorial planning is therefore conspicuously absent thus actions toward the application of African renaissance beyond conceptual frameworks are summarily torpid.

The economic and political reform mechanisms of NEPAD initiative of African renaissance theory are instructive. Concerning NEPAD economic reforms, besides seeking a peculiar economic basis of integration, the supportive activities of external development partners that usually slide into control are part of the legitimate fears that critics express. As for NEPAD political reforms, the three trends identified in the evaluation of democracy and good governance in Africa suffices.

This contribution opines that the anticipated African renaissance is a mindset and an outlook. As a mindset it is an educational process. There are two aspects of this process and they are focused on planning approaches. The first aspect deals with re-engineering epistemological foundations (Nabudere 2003). To this end there is a need to repackage the self-confidence of the average African and nurture their civic identity. African people should reconnect with those fundamental values of self-respect, dignity, pride, moral integrity, self-reliance and independence (Dembele 1998). With this the psyche of the African will echo the requisite Afro-centric outlook. Lots of African scholars, mostly published in CODESRIA documents, are following-up on this crusade although they focus more in making sensational pan-African statements than establishing theoretical foundations for change and developing realistic frameworks for concrete action.

The second aspect deals with the educational curriculum for planning studies. African planning education must of necessity retrospect into the history of African societies, their kingdoms and territorial configuration and their functional relationships especially in trade. This line of studies hopefully will address the background to the application of imperialism in the context of various phases of mercantilism in Africa. Imperialism impacts the space economy adversely. This explains current policy shift from urbanity to decentralization. The resultant pragmatic planning framework for sectoral interventions, which paralyses the place and function of land use change policies for managing space economy, is considered a move in the wrong direction. This trend is driven by the focus of attention on cross-cutting issues in planning such as urban environmental quality issues, which cause attention to focus on servicing the externalities of neoliberal economic activities. The core issues of planning, which are linked with reworking the space economy for the delivery of African renaissance, are not addressed. The current crusade for curriculum change subordinates the development of creative planning skills required for regional spatial integration. This is worrisome because the critical element of form and function in space economy is undermined.

At the moment the space economy in Africa is imperial that is to say it is extroverted and dependent on Euro-American mercantilism. The Euro-American mercantilism under reference highlights the clienteles' network in the development state, which sustains imperialism. Thus the resultant space economy disconnects with harnessing local resources and encourages the expansionist programs of neoliberal political economies. This explains the new agenda of sharing African market for Euro-American goods and services, which erupted at the turn of the twenty-first century. However the outlook of African renaissance, which contemplates territorial planning for regional spatial integration, draws from the need to introvert the space economy. Besides the challenge of reworking the space economy, the choice of the right planning instrument to execute the exercise is yet another challenge. The bottom line of the critical choice is growth although the point of departure is the determination of the nature of growth. The imperial growth, which Euro-American mercantilism is likely to deliver, is naturally not in contention.

8.2 Conclusion

The full realization of the futility in economic and socio-cultural development of the African region necessitated the philosophy of African renaissance. Therefore African renaissance is held in literature to be a philosophical and political movement to end the violence, elitism, corruption and poverty that seem to plague the African continent, and replace them with a more just and equitable order. As Africans, the time to act is now because in the words of Mangu (1998) 'we can no longer hide from the need to justify our existence, and reconcile ourselves with our people, our science and our history.' Progress made so far links with the NEPAD initiative. Although there is bound to be very strong dissensions, NEPAD is the clearest statement yet of Africa's determination to chart its own course of development while meeting internationally accepted principles of governance. This discourse concedes that the initiative is built on the principles of African renaissance as economic renewal strategy but given its globalist perspective it is obviously strapped with the apron strings of imperialism. Nevertheless, rejecting NEPAD at this time will amount to taking several steps backwards. The spirit of NEPAD alongside the political will of the African Union as well as the African Peer Review Mechanism (APRM) presents good opportunity for the way forward.

References

African Renaissance Institute (2000) The amended version of the vision, mission and objectives. Sandton, South Africa

Boloka GM (1999) African renaissance: A quest for (un)attainable past. Critical Arts 13(2):92–103

Botha T, Du P (2000) An African renaissance in the 21st century? Strategic Rev Southern African 22(1):1

Cossa JA (2009) African renaissance and globalization: a conceptual analysis. Ufahamu J African Stud 36(1)

Creff K (2004) Exploring ubuntu and the African renaissance: a conceptual study of servant leadership from an African perspective. Available at: http://www.regent.edu/acad/sls/publications/conference. Accessed 17 Apr 2015

Dembele DM (1998) Africa in the twenty-first century. 9th General Assembly CODESRIA Bulletin (ISSN0850-8712) 1:10–14

Van Kessel I (2001) In search of an African renaissance: an agenda for modernization, neo-traditionalism or Africanisation? Quest XV(1–2)

Makgoba MW (ed) (1999) African renaissance: the new struggle. Mafube & Tafelberg, Cape Town

Maloka T (2001) The South African "African renaissance" debate: a critique. African Institute of South Africa. Polis/R.C.S.P/C.P.S.R. 8 (*Numéro Spécial*)

Mangu AMB (1998) African Renaissance compromised as the dawn of the third millenium. 9th General Assembly CODESRIA Bulletin (ISSN 0850-8712) 1:14–22

Matunhu J (2011) A critique of modernization and dependency theories in Africa: critical assessment. African J Hist Cult 3(5):65–72

Muchie M (2004) A theory of an Africa as a unification nation: a re-thinking of the structural transformation of Africa. African Sociol Rev 8(2):136–179

Nabudere DW (2001) The African renaissance in the age of globalization. African J Political Sci 6 (2):11–27

Nabudere DW (2003) Towards a new model of production—an alternative to NEPAD. In: 14th Biennial Congress of AAPS. Durban: South Africa. www.mpai.ac.ug. Accessed 8 July 2011

Ndu Y (2014) The African renaissance in the 21st century: the challenges and the prospects. Available at: http://www.modernghana.com/news/565435/1/the-african-renaissance-in-the-21st-century-the-ch.html. Accessed 12 May 2015

Nmehielle VO (2003) The African Union and African renaissance: a new era for human rights protection in Africa? Singapore J Int Comp Law 7:412–446

Oelofse J (2013) African renaissance: African rebirth or African enslavement? Available at: http://globalreboot.org/2013/09/03/african-renaissance-african-rebirth-or-african-enslavement/. Accessed 12 May 2015

Ramose M (2007) In Memoriam. Griffith Law Rev 16(2)

Schwabe CA (2001) African renaissance: towards a spatial information system for poverty reduction and socio-economic development in Africa. (Paper presented to the 5th Africa GIS Conference Nairobi, Kenya 5–9 November)

Terreblanche S (2012) Lost in transformation. KMM Review Publishing Company, Sandton

United Nations Environmental Programme (UNEP) (1999) Global Environment Outlook 2000. Earthscan Publications, London

Vale P, Maseko S (1998) South Africa and the African renaissance. Foundation for Global Dialogue Occasional Paper, no 17, October

Van Niekerk AA (1999) The African renaissance: lessons from a predecessor. Critical Arts 13 (2):66–80

Chapter 9
Neo-Mercantilism as Development Ideology (in Africa)

Abstract Originally, neo-mercantilism emerged as a trade strategy, which is applied at the global space. The African region was exposed to it from the receiving end as a consumer economy. However its attribute of protectionism amongst other qualities recommend it as option to be conceptualized as development ideology for Africa. Neo-mercantilism as development ideology adopts government and entrepreneurial synergy to maximize regional interest as a means of advancement in political economy. Therefore it anticipates optimal state intervention to support entrepreneurship, both formal and informal although it holds informality as an exception and not a norm. In the circumstance, market force is expected to intersperse with planning rationality to manage growth in the context of regional spatial integration. Therefore, spatial factors framed on distributive justice sought with territorial planning principles, inform neo-mercantilist ideology as policy instrument for African regionalism.

Keywords Mercantilism · Neo-mercantilism · Century · Development ideology · Institutionalist · Developmentalism

9.1 Neo-Mercantilism: A Rationale for Change

Mercantilism is the political economic philosophy of the sixteenth through eighteenth centuries. It is characterized by the desire of nations to enrich themselves through control of trade. It emerged as a system for managing economic growth through international trade as feudalism became incapable of regulating the new methods of production and distribution. It was a form of merchant capitalism relying on protectionism. Hence, the concept of mercantilism covers protectionist policies to promote national economic development (Hettne 2014: 210). European nation-states used it to enrich their own countries by encouraging exports and limiting imports. This has the implication of extending market-space economy.

Epistemologically, mercantilism is a major determinant of urban form in Africa. Roughly three shades of mercantilism are discerned; namely, early mercantilism (tenth–fifteenth century), imperial mid-twentieth century European mercantilism covering the capitalist colonial period and the current twenty-first century Euro-American mercantilism and perhaps Chinese mercantilism under neoliberal dispensation in post-colonial period. The first the second shades were interspersed with slave trade period (fifteenth–mid-nineteenth century), which witnessed a very unholy form of mercantilism that elicits reparation. Nevertheless, mercantilism as an economic entity is contextualized by three schools of thought. Each school, including the reformist, institutionalist and neo-Marxist schools hold different positions. The position held by the institutionalist school seems to be predominant and inputs from this school are subtly regarded as orthodoxy in planning literature. The institutionalist school is more inclined towards a neo-liberal interpretation of trends since mid-twentieth century when Africa came under colonial rule.

The mercantilist era (tenth–fifteenth century) in the history of African civilization when resource management was under the control of charismatic leadership is recalled. At this period space economy expressed in the spatial segregation of resource and production areas guided the growth of Kingdoms and Empires. Then mercantilism was portrayed as pre-liberal economic policy in which nations or kingdoms are seen by mercantilists as large-scale versions of a private household, rather than as firms. The internet material which expressed this view also indicates that mercantilists share with modern neoliberals the 'view of world trade as a competition between nation-sized units'. It further indicates that in 'neo-mercantilist ideology the policies are national policies, directed ultimately at the welfare of the nation and not of the market'.

Classic mercantilism lost favour when the rising bourgeoisie grew tired of the limitations mercantilism placed on their actions. Liberating themselves required them to secure the policy of limited state action, which was termed 'laissez-faire'. In the 1970s, the bourgeoisie turned against post-war social Keynesianism because, as with mercantilism, it limited their actions and they embraced the ideas of Milton Friedman. Following Friedman's Monetarist ideas, state action remained but very much in the background skewing the market in favour of the wealthy (Anarcho 2005). The resultant predictable and predicted massive economic collapse, which empowered the wealthy, caused social Keynesianism to be replaced with neoliberalism as the de facto state principle.

With neoliberalism in place, the scale of space for mercantilism changed (see Seers 1983 cited in Hettne 2014: 223). Mercantilism seized to act as trade strategy on national space and extended to act on global space. The prefix 'neo' was added to mercantilism due to change in emphasis from classical mercantilism on military development to economic development, and its acceptance of a greater level of market determination of prices internally than was true of classical mercantilism. In essence, neo-mercantilism was founded on the control of capital movement and discouraging of domestic consumption as a means of increasing foreign reserves and promoting capital development. This involves protectionism on a host of levels: protection of domestic producers, discouraging of consumer imports, structural barriers to prevent entry of foreign companies into domestic markets, manipulation of the currency value against foreign currencies and limitations on foreign ownership of domestic corporations. The purpose is to develop export markets to developed countries, and selectively acquire strategic capital, while keeping ownership of the asset base in domestic hands. It therefore suggests new form of protectionism that is; qualitatively different from the traditional mercantilist concern with state building and national power that is the pursuit of statism.

Therefore, neo-mercantilism as a concept is seen as a policy regime that encourages exports, discourages imports, controls capital movement and centralizes currency decisions in the hands of a central government. This means the pursuit of economic policies and institutional arrangements, which see net external surpluses as a crucial source of profits. As an economic theory, neo-mercantilism maximizes benefits to the interests of a country such as higher prices for goods traded abroad, price stability, stability of supply and expansion of exports with concomitant reduction of imports.

Policy recommendations of neo-mercantilism are generally protectionist measures in the form of high tariffs and other import restrictions to protect domestic industries. This is combined with government intervention to promote industrial growth, especially manufacturing. This is why Raza (2007: 1) indicates that the concept is usually not publicly acknowledged by public policymakers although on account of these measures, China, Japan and Singapore are described as neo-mercantilist. Ironically, these same measures are currently responsible for bilateral trade relations rocking developed countries, sometimes eliciting scratchy comments quoted in Steven Schlossstein (1984). Nonetheless, market economists have argued the pros and cons of protectionism. The language of neo-mercantilist policies repeats the claims in earlier centuries that protective measures benefit the nation as a whole, and governmental intervention secures the 'wealth of the nation' for future generations. Indeed, the historical evidence leads any unbiased researcher to conclude that mercantilism has generally been successful in fostering economic development. Free-trade advocates have failed to muster counter arguments for why Britain fell behind the United States and Germany by 1880, after she abandoned mercantilism in favour of free-trade in the middle nineteenth century.

As it were, neo-mercantilism (appropriately referring to modern-mercantilism) retains most of the attributes of its parent notion (that is mercantilism) especially with regards to its concern for nation building rather than the inherent individual profitability attribute of modern-capitalism. Its concern for nation building separates it from the ethos of neo-liberalism and globalization. Neo-mercantilism therefore begets a peculiar neo-mercantile development ideology in which policies are national policies, directed ultimately at the welfare of the nation and not of the market, thus signalling protectionism. This presupposes mind-set and outlook issues determined by the status of productivity structures and levels of vulnerability in the global economy.

The logic for neo-mercantile development ideology for Africa is not farfetched. A recurring phenomenon in the history of African civilization is the development of market towns. Mercantile towns cradled African civilization followed by market towns termed colonial towns and then new towns that serve as administrative-cum-commercial hubs such as Abuja in Nigeria, Dodoma in Tanzania, etc. Most African cities remain largely commercial centers for trading in survivalist informal sector. The economy of most African countries maintain close relations with resource as in most east African countries or petro-chemicals as in Nigeria, or mining as in Angola, and so on. Therefore the ideological change is strategic on grounds of relevance, adaptability, comparative advantage in operationalization, and patriotism. Moreover the cost of neoliberalism reinforces the rationale for change.

Given the history of cities development in Africa, this discourse accepts the argument that robust marketing makes a city. To this end neo-mercantilism as development ideology for Africa enjoys some measure of justification. Africa is likely to remain a resource marketer into the foreseeable future considering its lack of technology to harness its resources and on the other hand the impact of global capitalist system reinforced through neoliberal economic policies to retain Africa in peripheral economy as source region. However, source region in the neo-mercantile ideology being expounded is conceptually not limited to the source of natural resources because transitional growth according to Rostow (1977) five-stage development of spatial systems is anticipated. This is why under neo-mercantile development ideology it is anticipated that African spatial systems can retain its market region outlook but not in the sense of neo-liberal permutations for purposes of extending the market-space economy of external economies. To the contrary, the market region legacy shall revert to its mercantile period significance for nation building in Africa.

Preference for neo-mercantilism is a strategic choice. The choice for neo-mercantilism draws from its concern for nation building coupled with its antecedents in nurturing traditional African civilization successfully. The neo-African planning theory it prospects to support shall assume the status of a general theory committed to handling humanistic interventions responsible for urban change. This outlook is mindful of population growth and urbanization and urban growth phenomenon anticipated in the African region in the new millennium. Moreover human-induced land use changes are considered the prime agents of global environmental changes (Ramachandra et al. 2012). Meanwhile the new planning theory engages spatial metrics analysis to address spatial equilibrium in planning alongside spatial determinism in economics. This elicits neo-mercantile planning paradigm which is built on five cannons; first the innovation of time element in planning, second upholding humanistic intervention as principal determinant of urban change, third merging economic and spatial planning, forth adopting creative outlook and fifth positioning transportation as a central element in spatial planning. Mindset and outlook issues are mainstreamed in the set-up.

9.2 Neo-Mercantilism as Development Ideology (for Africa)

In the effort to put forward neo-mercantilism as a model of development ideology for Africa, the perception of two terms is critical. These terms are 'ideology' and 'development'. The term 'ideology' is held to be a comprehensive normative vision, a way of looking at things, as argued in several philosophical tendencies (for example political ideologies). According to information in an internet post, recent analysis posits 'ideology' to mean 'a coherent system of ideas, relying upon a few basic assumptions about reality that may or may not have any factual basis'. It goes further to state that 'ideas become an ideology (that is, become coherent, repeated patterns) through the subjective ongoing choices that people make, serving as the seed around which further thought grows'. This definition, which accords with definitions such as given by Paul and Manfred (2010), suffices.

The other term 'development' is commonly associated with economic growth and modernization. Friedmann (1972) gave this indication although its usage is generally linked with economists. Friedmann's perception is applicable but its modernization attribute has imperial connotations. Graphically and more importantly, 'development' is held to be a growth process concerned less with modernization than with the spread of social justice, and the essence of social justice is not wealth but fairness. This position agrees with Takoma's (2013) postulation that industrialization, electoral democracy, and economic expansion are not things to be valued in themselves, but means (or, in some cases, obstacles) to deeper ends of social justice. Accepting Takoma's insight, the amalgamation of the terms 'development' and 'ideology' that is 'development ideology' is preferably perceived as a compendium of rationalized policies and aspirations conceived in the context of competing doctrines.

At the outset, neo-mercantilism as a development ideology draws from the conceptual meaning of mercantilism as trade methodology which Werner Raza (2007) presented. Hence its adaptation as a development ideology requires a theoretical foundation. The developmentalism theory provides this foundation in so far as the theory conforms to the notion, taken from an internet post, that the best way for Third World countries to develop is through fostering a strong and varied internal market, and perhaps to impose reasonable tariffs on imported goods. This notion is central to planning for regional spatial integration in Africa. Some other points of departure presupposes that developmentalism sheds its Eurocentric viewpoint of development, a viewpoint that often goes hand in hand with the implication that non-European societies are underdeveloped. It also has to shed its Universalist approach to development, which is not time and space specific. Its use of classic western standards, theories, and models including all appearances of vulnerability to imperialism and tendencies of neocolonial financial mechanisms will have to discontinue.

Neo-mercantilism as development ideology adopts government and entrepreneurial synergy to maximize regional interest as a means of advancement in political economy. Therefore it anticipates optimal state intervention to support entrepreneurship, both formal and informal although it holds informality as an exception and not a norm. In the circumstance market force is expected to intersperse with planning rationality to manage growth in the context of regional spatial integration. Therefore, spatial factors framed on distributive justice sought with territorial planning principles, will inform neo-mercantilist ideology as policy instrument for African regionalism. This is underpinned in the regional framework the Economic Commission for Africa (ECA) seeks for Africa.

In anticipation of new regionalism, neo-mercantilist ideology implicates the perception of functional regions as positive space economy. Earlier submissions on positive space economy suffice and are here correlated to neo-mercantile development ideology as a planning methodology. This will be elaborated shortly. Meanwhile, neo-mercantile development ideology also implicates worldwide cities concept. This is a new concept of cities classification that applies differently from the global cities concept and is presumed for use in the neoliberal frame of new regionalism. Therefore, mindful of global tendencies towards new regionalism there are three cognate scenarios in the conceptualization of neo-mercantilism as development ideology. They are: the intuitive platform, the socio-economic policy permutations, and the outlook of cities in neo-mercantile dispensation.

In the first instance, the intuitive platform signifies that the ideology requires a positive mindset and a visionary outlook which Mangu (1998) shared in his call of Africa to action. The requisite mindset takes bearing from the tenets of *negritude*, the cultural movement Davison et al. (1966: 22) referred to that portrays the rich heritage of African ancestry rediscovered by intellectual explorers to rescue a main section of humanity from unhappy misunderstanding. The mindset built on the threshold of negritude, hopefully will redress *a priori* colonial mentality bequeathed by colonialism and will help to repackage African civilization that will once again rest 'upon social and cultural advances of great antiquity' (Davison et al. 1966; 22).

Secondly, in tune with Demba Moussa Dembele's (1998) call to reconnect the African people with those fundamental values of self-respect, dignity, pride, moral integrity, self-reliance and independence, traditional ideologies such as *Ujamaa*, *Ubuntu* and *Omenani* will be used to define neo-mercantile development ideology as a socialization process. To this end, socialization into new values will be encouraged at homes and at civic and religious institutions and high fidelity compliance is expected given the malleable nature of traditional and contemporary African societies. The new values seek eco-centrism as standard for mobilizing the culture of eco-cities to serve as antidote, for what Simone's (1998: 17) identified as failed simulations of external notions of urbanity in African civilization. The new order preaches the values of reverence, humility, responsibility and care. The ethos of transcendentalism draws from these values as a transformative mechanism to encourage back-to-land measures but not in the traditional sense of subsistence. Indeed, a sense of starting all over within limits of existing asset base and conformity to affordable standard of living is in the bargain. This is intended to limit dependency and downsize the swag of international financial mechanisms in the development of the economic system.

The primary economic policy of neo-mercantilism as a development ideology rests on regional cooperation for resource marketing and agro-business. This is contemplated under the banner of NEPAD in its African renaissance frame. The policy provision implicates rigorous land system and land management reforms similar to the reforms contained in the Wise Land Use Act 2012 in South Africa. It is also geared towards integration that promotes new regionalism as envisaged in the Regional Integration Facilitation Forum (RIFF). However, following insights from Jon Woronoff's (1984) record of international reactions to Japan's surplus trade with the rest of the world, resource marketing policy and trade relations are built on the principle of fair trade. This is to say that the volume of trade transactions as two-way traffic will depend on the resource base of each African country. Countries with limited resources will moderate and downsize their trade transactions accordingly. This is to be expected anyway, but the hard truth is that this scenario carries along with it the lowering of standard of living in such countries. Affected countries, after a peer review evaluation, will be justified to engage in explicit protectionism regardless of globalization. Elevation from lower standard of living will be determined internally, pending the resourcefulness of such countries to grow their economy with their human and land resource.

Therefore, the philosophy of neo-mercantilism as development ideology holds that a positive space economy is a prerequisite for growth and development, on the condition that the space economy is free from distortions as speculated in Hicks (1998) submission, and resistant to Wallerstein's (2004) dependency thesis. The ideology also holds that the meaning of integrated urban economy follows Edwardo (1990) explanation of urban systems and is therefore incomplete if it is not expressed in space. This explains why the outlook of the ideology as a planning methodology accepts planning rationality as a requirement for spatial integration and growth. Thus, the ideology in principle deviates marginally from the creative attributes of formal planning theory. Thus, a neo-mercantile planning theory that draws from Baeten's (2012) argument to uphold creative principles in planning, and agrees with Simone's (1998) quest to pursue economic and sociocultural renaissance, suffices. And as entry point to African renaissance, the neo-mercantile planning theory commits to (re)think the space economy. This aspiration accords with Nabudere's (2003) thinking of re-engineering epistemological foundations of imperialism. Within the transition period, informality will be treated in neo-mercantile planning theory as an exception and not a norm but all within the framework of integration planning. To this end, Watson's (2009) work comes handy to program palliative measures, which trade-offs with remedial structures built with the anticipated neo-mercantile planning theory.

Thirdly, neo-mercantilism as development ideology shares the outlook that contemplates the worldwide cities concept for greater Africa. The 'worldwide cities' concept is a pathfinder perspective of global cities proposed for application in the delivery of new regionalism, in the context of neo-mercantile development ideology. The worldwide cities concept basically seeks synergy between Africans in Diaspora and Africans in the homeland. Therefore, the concept will associate with the 'hyperspace' syndrome that downplays spatiality, and upholds social equity across ethnic (racial) boundaries. Lessons from the worldwide conception of cities that sustain ecclesiastical states of Christian and Islamic kingdoms such as the Vatican City and the cities of Mecca and Medina, illustrate the space less limits of these cities that immortalize civilizations. Thus, it will realign the forces of Diaspora and bring the contributions of Africans in Diaspora to bear on the economy of greater Africa and in the process, rework the status of Africa in the world system. A synonymous scenario relates to the functioning of Jewish Americans in relation to the State of Israel. These cities will be propelled with the instrumentality of human culture that subordinates all other development factors in science and technology. The conceptual city will become operational with the slogan 'The African city is Here' implying that the individual African is an embodiment of his homeland city, essentially the urban region that is personified where-so-ever he finds himself, anytime.

The main challenges before neo-mercantile development ideology are mainly in the domain of epistemological ideologies such as imperialism, the informal sector, in-formalization, informal planning, liberalization vis-à-vis free-trade and anti-protectionism, etc. These ideologies are critical support structures that used to establish the hegemonic influence of neoliberalism. And for the hegemonic status of neoliberalism to remain, Leys (1990, quoted in Peck et al. 2009) imputes that: 'it is merely necessary that it has no serious rival.' So to rescue Africa from the imperial straps of neo-liberalism will require commitment to experimentation, and as Simone (1998: 108) puts it, 'experimentation is not risk free'. Primary amongst the straps that require serious attention is funding. To this end, neo-mercantile development ideology is conceived with some measure of time-bound socialist principles to encourage minimalist funding, and exploit reliance on human capital for harnessing agrarian agriculture as leeway for agro-business.

Meanwhile neo-mercantilism is framed with a scale of space and time that is to say, it is meant for Africa in the twenty-first century. Within this period, the growth vision for Africa is framed to secure food sufficiency and security. Beyond the twenty-first century the tenets of neo-mercantile development ideology is due for review but hopefully then, Africa will be properly positioned to engage global economic orthodoxies.

9.3 Neo-Mercantile Versus Neoliberal Development Ideologies

There are impressions in literature that neoliberalism is conceptually synonymous with neo-mercantilism because both concepts support protectionism in different forms. To establish this critical observation, the link between mercantilism and capitalism is recalled. For mercantilism, state action was an essential feature, used to bolster the ruling elite by pursuing policies which increased their power on the market. In neo-mercantilism, the scenario is not markedly different except for outlook issues which tilt towards pursuing regional rather than national growth interest in the global economy. On the other hand, the policy of limited state action in neoliberalism has the effect of maintaining and protecting capitalist property rights and skewing the labour market in favour of the bourgeoisie. The central strategy in both concepts is protectionist and in favour of the elites. The difference between the two concepts comes alive with contextualizing them as development ideologies. This being the case, it is most likely that different tactic will evolve to retain protectionism not as a matter of convention but in pursuit of substantive values (objectives), which in the case of neo-mercantilism transcends private profitability.

The bone of contention between the two ideologies lay in the antithetical posture of neoliberalism towards African renaissance and its vulnerability to neocolonial tendencies. The issue of informality that characterizes neoliberalism and neoliberal planning theory, its instrumentality for the expansionist program of external economies and the presumed perceptions of Africa as consumer economy are matters of critical concern. The same concern is expressed for the extension of market-space economy through free trade in neoliberal context in which weak states are at the receiving end.

The major points of departure between neo-mercantilism and neoliberalism as development ideologies are in their planning methodology, their philosophy, their motivation, and their determinant factors, particularly liberal market force and financial mechanisms. Neo-mercantilism and neoliberalism as development ideologies share discretely the ethos of international trade vis-à-vis fair trade anyway, privatization, social stratification and globalization. As planning methodologies, both ideologies share participatory process. But for neo-mercantile planning, the process is within limits of democratizing and not liberalizing planning decision as it is the case with neoliberal planning methodology.

The philosophy of both ideologies share the ethos of integration but neo-mercantile ideology emphasizes integration in geographic space, and not in abstract economic space as is the case with neoliberal ideology. By implication, neoliberal ideology seeks the economic bases of integration with neoclassical theories, while neo-mercantile ideology pioneers the search for spatio-physical bases of integration which places emphasis on space economy viewed from planning perspective. Positive space economy is therefore critical for neo-mercantilism as instrument to address national development and to a minimal extent, private profitability in neoliberal terms. Ultimately, it is incumbent on neo-mercantilism to give considerable attention to prudent use of space that is subsumed in planning rationality. The reverse is the case for neoliberal ideology in which market force prevails in total disregard for planning rationality. Thus the defining moment of divergence is in determining the requisite planning theory for the delivery of the different development ideologies.

Both ideologies are receptive to new regionalism but from different perspectives. Neo-mercantile ideology is mindful of the resilience of epistemological ideologies that derived from global regionalism in Africa since mid-fifteenth century. New regionalism must be seen to be free from these epistemologies particularly imperialism, dependency, informality and declining productivity. Incidentally, the prospects of discontinuing with these epistemologies is not clear, given the current global political economy in which the expansionist program for market-space economy informs the new scramble for Africa.

The neo-mercantilist ideology rejects neoliberal reforms such as tax cuts for corporations and the wealthy, the abandonment of capital controls, the removal of democratic controls over central banks and monetary policy, and the deregulation of financial industries and so forth (Briggs and Yeboah 2001; Brenner and Theodore 2002). These reforms do not fall within the pro-Keynesian and pseudo-capitalist economy anticipated for neo-mercantilist ideology. Developed countries follow pro-Keynesian polities in practice and mobilize neoliberal prudent economists to developing countries, who insist that developing countries should not also pursue the same stimulus economic development principles (Enwegbara 2015). According to Enwegbara, 'the hidden agenda here is the fear that allowing developing countries pursue the same pro-Keynesian policies, would make them become industrial powers too'. He leaves the impression that anti-Keynesian macroeconomic policies across developing countries are promoted by 'local neoliberal economists and those educated and nurtured in western controlled multilateral institutions like IMF and the World Bank'.

Thus the challenges before neo-mercantile development ideology are mainly issues in the domain of epistemological ideologies. These ideologies are contained in the repressive attitude towards formal planning vis-à-vis master planning, the in-formalization process vis-à-vis informality in planning, the free-trade syndrome, and the repressive attitude towards protectionism. Neo-mercantilism is framed with a scale of space and time that is to say it is meant for Africa in the twenty-first century. Therefore, it is not answerable to the epistemologies identified. The greatest threat is funding although neo-mercantile development ideology is conceived with some measure of socialist principles to encourage minimalist funding and exploit reliance on human capital (given urbanization indictors) for agrarian agriculture en-route agro-business.

Indeed the relationship between neoliberalism and neo-mercantilism as development ideologies elicits concern for mutual respect otherwise systemic undermining may result with Africa at the receiving end. This scenario should be avoided although the Chinese communist Cultural Revolution stood up successfully against capitalism. The hegemonic outlook of neoliberalism has to respect the right of Africa to chart her destiny with neo-mercantile development ideology and be prepared to co-operate with neo-mercantilism at their points of interface. Therefore, the interface epitomized in protectionism will determine bilateral relations for Africa in the global economy. An interactive forum of regional stakeholders will continually rework the ground rules for interactions, however, in such a way that guarantees the local control of African market for the twenty-first century global mercantilism.

9.4 Conclusion

The wisdom to conceptualize neo-mercantilism as development ideology for Africa is not necessarily built on institutionalist school of thought. The trend of events built on this school of thought does not appear to be favourable to the delivery of African renaissance of the twenty-first century. An alternative school of thought

9.4 Conclusion

that is inclined to the position of neo-Marxist school suffices. It anticipates the transition to neo-mercantile development ideology in which planning theory shall re-engineer mainly through a natural process. In practical terms, because planning theory with capacity for regional integration is envisaged, it shall highlight formal planning attributes. Participatory process will be mainstreamed more as a political necessity and as a missing aspect of plan implementation than planning requirement. Indeed the fundamental challenge ahead of planning scholarship lies with the difficulty of combining economic and spatial planning (see Parker et al. 2001). But this can be handled by elevating the status of spatial planning to a position where it occupies overarching influence over all other forms of development planning. Hitherto the reverse is more-or-less the case. Sectoral and project planning should relocate to merge with plan implementation and budgeting procedures where they will be subject to the design functions of spatial planning.

References

Anarcho (2005) Neoliberalism as the new mercantilism. Available at http://www.anarkismo.net/article/192. Accessed 17th Feb 2015
Baeten G (2012) Neoliberal planning: does it really exist? In Taşan-Kok T, Baeten G (eds) Contradictions of neoliberal planning. Department of Social and Economic Geography, Lund University, 22362 Lund, Sweden pp 206–210
Brenner N, Theodore N (2002) Cities and the geographies of 'Actually Existing Neoliberalism'. Antipode 33:349–379
Briggs J, Yeboah IEA (2001) Structural adjustment and the contemporary sub-Saharan African city. Area 33(1):18–26
Davison B, The Editors of Time-Life Books (1966) Africa kingdoms, vol 37. Time-Life Books, New York
Dembele DM (1998) Africa in the twenty-first century. In: 9th General Assembly CODESRIA Bulletin (ISSN 0850-8712) 1:10–14
Edwardo LE (1990) Community design and the culture of cities. Cambridge University press, Cambridge, p 35, 38, 68, 90, 91, 109, 143, 154 (356p)
Enwegbara O (2015) Is this the end of neoliberalism in Nigeria? Punch Newspaper. Apr 16
Friedmann J (1972) A general theory of polarized development. In: Hansen NM (ed) Growth centers in regional economic development. Free Press, New York
Hettne B (2014) Neo-mercantilism: what's in a word. Available at: http://rossy.ruc.dk/ojs/index.php/ocpa/article/viewFile/3924/2090, p 205–229. Accessed 4 Mar 2015
Hicks J (1998) Enhancing the productivity of urban Africa. In: Proceedings of an International Conference Research Community for the Habitat Agenda. Linking Research and Policy for the Sustainability of Human Settlement held in Geneva, July 6–8
Mangu AMB (1998) African renaissance compromised at the dawn of the third millennium. In: 9th General Assembly CODESRIA Bulletin ISSN 0850-8712 (1):14–22
Nabudere DW (2003) Towards a new model of production—an alternative to NEPAD. In: 14th Biennial Congress of AAPS. Durban: South Africa. <www.mpai.ac.ug>. Accessed 8 July 2011
Parker DC, Evans TP, Meretsky V (2001) Measuring emergent properties of agent-based land use/land cover models using spatial metrics. In: Seventh annual conference of the international society for computational economics. Available at: http://php.indiana.edu/~dawparke/parker.pdf. Accessed Sept 2003 (cited in Herold et al 2005:375)

Paul J, Manfred S (2010) Globalization and culture, vol 4: Ideologies of globalism. Sage Publications, London

Peck J, Theodore N, Brenner N (2009) Postneoliberalism and its malcontents. Antipode 41(1):94–116

Ramachandra TV, Bharath HA, Sreekantha S (2012) Spatial metrics based landscape structure and dynamics assessment for an emerging Indian megalopolis. Int J Adv Res Artif Intell 1(1):48–57

Raza W (2007) European union trade politics: pursuit of neo-mercantilism in different arenas? In: Becker J, Blaas W (eds) Switching arenas in international trade negotiations. Ashgate, Aldershot

Rostow WW (1977). Regional change in the fifth Kondratieff upswing. In: Perry DC, Watkins AJ (eds) The rise of the Sunbelt cities. Sage, Beverly Hills

Schlossstein S (1984) Trade war: greed, power, and industrial policy on opposite sides of the pacific. Congdon and Weed, New York, p 4

Simone A (1998) Urban processes and change in Africa. CODESRIA Working Papers 3/97 Senegal, p 12

Takoma P (2013) Development as ideology. Available at: https://dartthrowingchimp.wordpress.com/2013/01/17/development-as-ideology/. Accessed 17 Feb 2015

Wallerstein I (2004) World-systems analysis: an introduction. Duke University Press, Durham

Watson V (2009) 'The planned city sweeps the poor away…':urban planning and 21st century urbanization. Progress Plan 72:153–193

Woronoff J (1984) World trade war. Praeger, New York, pp 144–145

Chapter 10
Introducing Neo-mercantile Planning Theory

Abstract The neo-mercantile planning theory supports the spatio-physical growth theories typology, developed along the convergence hypothesis model. It introduces the 'Time-efficient' effect (measured with a 'Time-coefficient') in planning, which classifies the time efficiency of cities to move goods and services in global space. It argues that the system of commercial nodes and their distribution in space is a function of time-efficient decision-making for demand-side trade transaction. Within neo-mercantile planning process sectoral and project planning relocates to merge with plan implementation procedures where they are subject to the provisions of spatial integration mechanisms. Also, budgeting is strategically subsumed in the neo-mercantile planning framework which has inbuilt provision for investment planning and funding strategies. In practical terms, the theory functions with an innovative spatial model for urban region development to translate growth visions of urban Africa into space. It does so with spatial integration planning (SIPs) and thematic integration planning (TIPs) instruments. Three levels of planning and six categories of spatial systems shall be integrated in line with the standards of extended metropolitan region and growth triangle models for isolated market regions.

Keywords Neo-African · Time-efficient · Time-coefficient · Neo-mercantile · Spatio-metric · Trade-basin

10.1 Neo-African Spatial Development Theory

The theoretical framework of spatial planning earlier identified, which is found to be oriented towards neoliberal planning principles, comprises 13 typologies of objectives. The reinforcement of this framework to enhance its capacity for integrative planning necessitates inputs in eight typologies of objective shown in Table 10.1 where research findings identified shortfall in principles of creative planning. Most of the findings gravitated under the rubric of planning methodology, followed by planning instruments and then the nature of planning. This distribution

Table 10.1 Matrix of new provisions for capacitating the theoretical framework of spatial and statutory planning in Africa

Objectives	Options	Criteria
The concept of urban. environment		
Statutory planning	• To improve the place and function of spatial development planning in statutory planning provisions. • Standardize the content of planning instruments	• Neo-African development ideology • Urban development strategy • Spatial design models • Urbanization ideology • Growth visions • Policy interpretation
Spatial planning	• To adopt spatial planning as a form-based concept for managing land use development.	• Spatial determinism • Spatial equilibrium • Urban growth management • Land use systems • 3-dimensional planning • Functional flow analysis in space • Urban design standards
Nature of planning	• Revision to visionary planning outlook and the adoption of territorial planning theory that brings form-based planning to bear in development planning systems	• Long-term structure plans • Detailed plans • Planned Unit Developments (PUDs) • Assets utility planning
Purpose of planning		
Planning instruments	• To adopt a thinking instrument for developing planning paradigm • Model an overarching neo-African spatial planning initiative as instrument for trans-national, national and local levels of territorial planning • Subject public sector infrastructure development to territorial planning approval	• Neo-mercantilism • Neo-African planning paradigm • Spatial Equilibrium Approach (SEA) • Merchant cities model

(continued)

10.1 Neo-African Spatial Development Theory

Table 10.1 (continued)

Objectives	Options	Criteria
Participatory process	• Retain participation at consultative level.	• Mainstreaming monitoring • Participatory development • Feedback mechanisms • Intensive participation
Planning methodology	• Standardize and establish statutory spatial planning methodology • Institute mechanisms for regular accreditation of public sector planning departments by professional registration councils • Establish interactive forum for researchers, consultants and policy makers to serve as quality control mechanism for planning consultancy services	• Spatio-metric planning model • Asset-based analysis • Multifunctional space analysis (time budget)
Urbanism		• Urban design theory
Planning knowledge		• Laws of informality
Plan evaluation technique		
Regional integration	• To adopt structural plans as a precondition for budgeting • To develop a visioning process • Institute vertical planning	• Urban market regions • Resource Management Plans • Growth vision—spatial model—road maps • Borderless worldwide cities concept • Intervention plan
Cross-cutting issues	• Unlock resistance to planning through engagement • Adopt visual presentation in the planning toolkit	• Cooperative governance • Spatial design concepts

Source Own construction (2013)

of findings was to be expected because activities in those areas of planning as expressed in the literature are highly contentious. The inclusion of the new provisions to the status quo validates the new theoretical framework as a working instrument for spatial planning in Africa. Where there is conflict(s) with reference to the emerging theoretical framework shown in Table 5.7 especially in the case of participatory planning provisions, the submissions hereunder supersedes.

The inputs identified drew from neo-African mindset that anticipates spatial regional integration. These inputs were chosen tactically to create a favourable environment for territorial planning. Planning traits hitherto subordinated based on neoliberal planning rationale were reconsidered. This is because the new

lanning, which neoliberalism suggests, is not likely to help Africa
on in spatial terms. What this means is that Africa's urban
mized in low productivity, which started mid-nineteenth century
perialism, is likely to continue.

10.1.1 Major Argument of Neo-African Spatial Development Theory

The neo-African spatial development theory argues for the introduction of time factor in land-use planning for economic growth. Time factor is the time required to reach inter or intra-trade locations in an urban region. This argument draws from the experiences that are contingent on the annals of Kingdom and Empire building in Africa especially during the mercantile period (tenth–fifteenth century). Again it draws from existing regional development theories although these theories have contributed little from spatio-physical perspective. This partly explains the reliance on historical experience that is scarcely theorized in which the time factor played a significant role in the resolution of trade relations. It also draws from the concept of urban distribution mentioned earlier, whereby people seek to improve their roles in the urban system by reducing the cost and time of overcoming distance (see Edwardo 1990: 38).

The time factor is therefore considered necessary for the synthesis of economic and spatial planning in contemporary spatial systems. It is an inevitable interface for creating utility from economic and spatial planning perspectives. Basic economic principles indicate that goods and services acquire utility when they are available at a particular place and at a particular time. On the other hand, land-use planning of popular design tradition in southern towns of forest zones in western Sudan distributes trade locations in space and time in a manner that prevents overlap. The time-related principles in economic and planning are still valid.

Mercantilism implies that trade as well as time is a factor of trade. Movement of goods, services and people for trading activities has always been time-specific. The history of African experience in this regard indicates that trading posts locate at neutral spots where it is time-efficient for neighbouring native settlements to attend, transact business and retire to their home base. Weber's (1929) location theory skipped the mechanism that determines the location of subcentres or markets in his explanation of the functional flow of market operations in spatial systems. But he recognizes that these subcentres exist for trade relations. The subcentres that occupy the neutral spots earlier mentioned evolve as explained in Rostow's (1977) stages of growth theory into commercial settlements. However they are traditionally differentiated from native home base settlements on grounds of civic identity and the exercise of authority over resource management. The elites in home base locations exercise such authority as explained in Friedmann's (1972) general theory.

Time-efficient access to trade locations in urban regions determines the productivity of trading posts as measured by turn-over in the volume of trading transactions. The phrase 'Time-efficient' is coined to define the quantum of time required to access goods and services in space. Face-to-face contact is implied irrespective of advances in communication systems (see Edwardo 1990). Structuralist theorists are yet to examine the effect of transaction turn-over in regional development notwithstanding the transaction cost introduced in new institutional economics in the 1970s to the 1990s periods. Suffice it to say that the system of commercial nodes and their distribution in space is a function of time-efficient decision-making for demand-side trade transactions. Tiebout (1956) and North (1956) export-base theory touched on this in their explanation of economic growth although their effort had little to do with the use of space.

Christaller (1933, 1966) places theory at the starting point, and explained the spatial segregation of manufacturing and service functions and later used marketing principles to establish central place patterns. It is not clear how his marketing principles relate to demand-side marketing. Nevertheless, the demand-side marketing cannot be denied its influence in the management of outlets for service functions which are more or less trade functions. The impact of the influence particularly sought the distribution of sales outlets in space. It is presumed that these outlets have comparative advantages which are traceable to their time-efficient properties. The time-efficient properties include location, proximity to population concentration, size of catchment area, intervening opportunity provision, accessibility, management (trade transaction) services, etc. From these properties time-efficient coefficient (which measures 'Time-efficient effect') for trading outlets can be determined. The coefficient measures the strategic position of a city to facilitate trade. It links with the opportunity cost of using spatial systems for trade relations. Polarizing behaviour could be factored on these coefficients to determine the hierarchy of settlements, their links and relationships and the attendant distributive networks. And time-efficient networks are presumed to attract more goods and services resulting in development corridors and growth triangles.

From these submissions the neo-African spatial development theory is hypothesized as a general theory for regional integration within spatial systems in Africa. It addresses the spatial bases of integration, unlike existing regional development theories which address the social and economic bases of integration. The neo-African spatial development theory pioneers neo-mercantile theories, a new set of theories next to neoclassical theories for regional development. Although through the consideration of geographic space and concern for the spatial dimension of regional growth and trade it links with neoclassical theories. Within neo-mercantile theories it belongs to the spatio-physical growth theories typology developed along the convergence hypothesis model. It shares the principles of endogenous growth theories of neoclassical theories because it accepts that the growth of the region is internal. In the subsequent section, the attributes of neo-mercantile planning theory is shared. It will be seen that its spatio-physical perspective is reflected in its conceptualization, approaches and frameworks. The relationships of the framework instruments and their spatial organization are modelled in a theoretically compelling manner.

10.1.2 Neo-mercantile Planning Concept

The proposed neo-African planning concept for delivering spatial planning activities is dependent on the political will to declare neo-mercantilism as a development ideology in Africa. The declaration which literally confirms the current status quo in national economies is used to theorize development for Africa. The proposed planning concept is technically termed neo-mercantile planning. The term neo-mercantile planning derives from combining neo-mercantilism and spatial planning. The resultant phrase literally means planning in spatial systems where the development ideology is determined by neo-mercantilism. In other words, it infers planning within spatial systems where mercantile economy prevails. In neo-mercantile planning, neo-mercantilism serves as thinking instrument as such neo-mercantile planning has three attributes that informs it as a planning concept. These attributes are: spatio-metric investigation, comprehensiveness and substantive orientation. Neo-mercantile planning concept is a scientific process of articulating the use of space to achieve growth visions and shape the city in Africa. Its rationale rests on the knowledge that pragmatic, short-term, process-oriented planning has consistently failed to source integration in the past three decades in Africa.

Spatio-metric investigation approach associated with neo-mercantile planning is an innovative technical methodology of studying urban growth. Spatial metrics was introduced in the mid-1980s in the literature of ecology and by late 1990s it was adopted in geography and landscape architecture for describing and comparing the structure and form of the various cities (see Poulicos et al. 2012: 263). The use of spatial metric concepts for the analysis of urban environments is starting to grow (see Herold et al. 2005: 375). Typical applications include an estimation of metrics to describe an urban environment with particular emphasis on the urban versus non-urban dichotomy and the computation of metrics for the same city or region for different time periods to assess the dynamics of change (see Poulicos et al. 2012: 263).

Spatial metrics is defined as measurements derived from the digital analysis of thematic-categorical maps exhibiting spatial heterogeneity at a specific scale and resolution (see Herold et al. 2005). Spatial metrics is used generally to analyse the spatial and temporal dynamics of urban growth. The metrics quantify the temporal and spatial properties of urban development, and show definitively the impacts of growth constraints imposed on expansion by topography and by local planning efforts (Herold et al. 2003). As a geospatial tool it has great potential for long-term monitoring and assessment of urban growth and its associated problems in surrounding land cover. Spatial metrics are found to provide the most important information for differentiating urban land uses and Herold et al. (2003) argue that the combined application of remote sensing and spatial metrics can provide more spatially consistent and detailed information on urban structure and change than either of these approaches used independently.

10.1 Neo-African Spatial Development Theory

The comprehensive planning attribute of neo-mercantile planning draws from the new orientation of comprehensive planning concept which is marked by a transition to policy analysis and emphasis on procedural change. It looks at the spatial dimension of all strategic policies and aims at integrating and coordinating all space-consuming activities in a geographic territory. It brings together and integrates policies for the development and use of land with other policies and programmes which influence the nature of places and how they function. The inherent procedural/methodological renaissance and scoping define the point of departure for the design of spatial frameworks in which the evolution of society and economy could be accommodated (Albers 1986: 22). This elicits increased attention in understanding local contexts and formulating guiding frameworks that inform land development and management (Landman 2004: 163). This slant of comprehensive planning is encouraged, mindful of the threat the transition to policy analysis poses to planning interventions in geographic spaces.

Neo-mercantile planning is a substantive planning concept. It is a planning methodology and a development ideology as it is found within the IDP initiative in South Africa. In other words, it is a planning instrument intended to address an approach to development. As found with the IDP concept, it is a hybrid concept, the objectives of planning make it a development ideology and the sequence of activities makes it a methodology. Its central objective is to secure productive urban forms. It perceives the urban form as an ecosystem with functional parts—land-use systems, distribution and patterns—that interact continually within spatial systems. As ecosystems are hierarchical and site-specific, so is neo-mercantile planning in its disposition towards regional integration. Hence it applies at the territorial levels of the urban core (local level), the urban economic region (municipal level), and the urban market regions up to trans-national levels. All the levels are unified in sourcing economic growth and productivity through spatial regional integration.

A neo-mercantile planning concept makes humanistic interventions in the environment the focal point of planning interventions. The human systems expressed in the distribution of population and activities in space and time is critical for neo-mercantile concept. The planning methodology retains people as principal elements of planning but not as instruments in the planning process. The methodology adopts a mixed approach to participation. Participation is consultative until the plan implementation stage when it transits to interactive participation.

Neo-mercantile planning offers a centralized system of managing decentralized planning. This study found that central control, either through directives or guidelines, is inevitable in so far as planning remains a public sector activity. It is even more so where the planning instrument is meant to serve as integrative instrument at continental level. Otherwise the integration sought will be defeated logically.

Neo-mercantile planning methodology comprises six stages of activities, namely; mobilization, population studies, existing condition studies, plan generation, integration (spatial design) and implementation techniques. Succinctly, mobilization deals with mindset and growth vision issues, population studies deal with assets analysis, existing studies deal with urban (settlement) form, plan generation deals with infrastructure projections, integration deals with design concept, and implementation deals with project planning, funding and monitoring.

In principle neo-mercantile planning is not committed to urban structure theories. Rather it identifies with the structure offered in the definition of the urban environment using the systems or spatial organization approach. The concept is also not committed to the core-periphery theory—rather it stands with the export-base theory and draws its arguments inevitably from Marxist theory. Hence, neo-mercantile planning shares visionary and territorial outlooks and to some extent a technocentric outlook. Neo-mercantile planning implies long-range restructuring and reintegration of the spatial structure of urban regions in Africa, involving long-term planning; however, phased in periods of short duration. Its territorial outlook informs its status as a form-based concept that studies and simulates relationships between land-use systems, distributions and patterns in three-dimensional space. It adopts form and function as overarching principle to address spatial integration in urban regions and ultimately impact the spatial structure of space economy. Therefore the distribution of activities in space and in time-efficient manner preoccupies neo-mercantile planning.

As a technocentric concept it is concerned with proper kinds of means. To this end it perceives the natural environment as natural matter from which man can profitably shape his destiny. Therefore, it accepts population pressure and economic growth as inevitable forces that can and must be accommodated by proper multiple purpose management and the application of sensitive planning checks and balances.

Neo-mercantile planning can be defined as assets utility planning, used to manage the use of space in cities and city regions. It seeks to deliver substantive spatial integration intended to secure productivity in the cities and urban regions and within limits of modern environmentalism. To that extent it is eco-centric and subject to environmental laws in deciding space-activity relationships. Hence it is concerned with ends that seek to achieve sustainable urban form which respects the integrity of the natural ecosystem and the cultural habitat. Alternatively, it can be defined as long-term visionary activity requiring multi-professional skills that act principally on health, culture and informal sector, transportation, and trade issues and regarding them as influences that determine the form of spatial systems.

The planning paradigms required to execute planning concepts are found to have problems with resolving issues of participation, the synthesis between the use of formal and informal expert knowledge, the interface between political and technical analysis, funding mechanisms and bridging the gap between principles and practice. How these issues will be handled in neo-mercantile planning is highlighted in the next sub-section.

10.1.3 Neo-mercantile Planning Paradigm

Neo-mercantile planning theory postulates an integration planning paradigm. This approach is concerned with the methodology of integration. Not all methods are acceptable. It was found in the study of the IDP initiative in South Africa that integration planning was corrupted—in which integration by stapling was engaged. The proposed integration planning paradigm combines the qualities of product and process-oriented planning to provide form-based planning for the development of spatial systems. In this way, it establishes frameworks for the spatial organization of activity systems by establishing and assigning functions to activity belts for service, manufacturing and primary activities. Activity belts relate to zoning principles but not rigidly, because permissible levels of heterogeneous land-use patterns are allowed. Recourse to zoning principles is justified to sanitize the bizarre land-use patterns which developed under neoliberal planning dispensation, partly leading to the informalization of African cities. Bibangambah (1992) described it as the progressive 'ruralization' of urban areas in Africa. The activity-belts principle is used to achieve a spread-effect in spatio-physical terms for project development activities and this has implications for urban form. This principle will be supported with urban growth management instruments such as urban growth boundaries, urban development boundaries, urban edge, etc.

The integrated planning paradigm upholds form and function principles. This principle is established in planning to support economic growth and productivity. It correlates with the time-efficient argument for trade relations and activity-belt principle of functional flow relationships. Form at the urban level relates to urbanity and at the regional level to spatial organization.

The integration planning paradigm adopts renewed product-planning principles which are characterized by a shared vision and long-term objectives that inform a draft plan. This incorporates participatory principles which apply as fortuitous circumstance permits in the planning process. What this implies is that participation is not universal and not always interactive. The creative process is therefore secured from quackery and charlatanism and even usurpation. The draft plan which is being protected provides the spatial guidelines for infrastructural and land-use development. The transition of draft plans into space which proceeds in line with participatory planning principles requires of the planning system to retire and serve as quality guides, more of refereeing job than participatory. This should not be likened to the current development control procedure because an integrated system of public, organized civil society and judiciary monitors will be engaged to manage development control. In other words, the authority to control development will be subject to participatory principles. The monitoring system is authorized to stop spurious development permits being issued by the planning apparatus.

The integration planning paradigm seeks dynamic equilibrium in the synthesis of political and technical analysis in planning. For the creative aspect of integrated planning technical analysis prevails and vice versa for intervention procedures; however, within the limits of decisions reached in the creative segment. Overall formal expertise knowledge coordinates planning inputs in consultation with informal expertise knowledge inputs which are channelled through organized civil society. Intensive consultation is intended. Politicians will have to be engaged extensively in the creative process because it is found that politicians and not technocrats control decisions in budgetary allocations and it is politicians who exercise authority in managing development in the kind of weak states found in Africa. This approach, as is found in this study and stated earlier, has potentials to unlock resistance not antagonistically in terms of the actors that planning has to engage.

Integration planning integrates funding mechanisms in its operations and prospects to serve as referral input with some measure of authority in national economic development planning. In essence economic planning, budgeting and spatial planning are horizontally integrated at all administrative levels. Segregated decision-making is discouraged. Hitherto spatial planning has occupied the rear position in segregated decision-making systems and this is found to adversely impact the authority of spatial planning in governance. Therefore integrated planning provides the platform for budgeting and project development. It functions as an overarching planning instrument with sectorial and project planning subsets within it. The activity belts will serve as control to manage the location of sectorial projects in space. In this way, the location of projects in space is allowed some measure of flexibility required for the involvement of external development partners.

The theoretical framework for the integration planning paradigm is discussed in the next section. Provisions in the framework are made to ensure that strategies employed to overcome identified problems of existing paradigms are effective. The theoretical framework flows with current trends in planning; however, this is done discreetly and provisions are in place to ensure that visionary and form-based planning structures are secured.

10.1.4 Neo-mercantile Planning Theoretical Framework

The reality of development processes varies amongst nations, thus requiring variations in the instruments for planning interventions. In this research, attention is focused on scenarios with abundant natural and human resources that reveal predicaments in the form of declining urban productivity linked with inadequate development of the space economy. Such scenarios exhibit spatial distortions in the development of the urban regions that are left unresolved in pursuit of market economy in neoliberal terms. The resultant extroverted economy prevails in the context of disconnect between the urban and rural economies and the proliferation of a survivalist informal sector economy. This scenario is common with the so-called source regions in global economy and their revision, wherever it is found, especially in the African context, provides the rationale for the neo-mercantile planning paradigm.

The application of a neo-mercantile planning paradigm requires a mindset and an outlook with the objective of spatial regional integration. The mindset that can mobilize political will is required and this can most effectively be informed through world systems analysis of trends in global development ideologies. Given the depth of structures that sustain the status quo the outlook of interventions must be visionary and this has implications for requisite thinking instrument(s). The consideration given to neo-mercantilist ideology to serve as thinking instrument is based on the fact that neo-mercantilism complements neoliberalism in proffering globalization but the concern of neo-mercantilist ideology is directed ultimately at the welfare of the nation and not of the market, thus signalling protectionism. Hence it fits very well as a platform to design instruments for the delivery of spatial regional integration which eluded neoliberal planning paradigm (instrument). Although in the participatory process suggested in neoliberal planning, there was no theoretical base for spatial integration.

The theoretical framework for neo-mercantile paradigm upholds a visionary planning outlook with the capacity to redress spatial distortions in the urban regions. This implies the use of planning rationales to address land-use management in a sustainable manner. Hence the framework supports the principles of form and function in planning delivered with formal expertise knowledge. Participation in the planning process will be limited in line with research findings. The framework facilitates form-based planning, hence its use of provisional planning instruments. These characteristic traits are found to be subjugated in favour of politically motivated broad-based guidelines and the use of market forces to facilitate project planning. This created room for organizational planning and informal planning as well as manipulative participation in the planning process. These provisions were down-sized in the review of the status quo in theoretical frameworks for the integration planning paradigm.

The theoretical framework for neo-mercantile planning as contained in Table 10.2 is made compliant to form-based planning as the research finding suggests, and especially for purposes of achieving spatial regional integration. The framework provides immunity for functional planning instruments and respects participatory process in plan implementation without compromising formal planning procedures. Unfortunately the liberalization of planning decision is not considered to be part of the working conditions of the neo-mercantile paradigm notwithstanding the inputs craved from politicians.

The theoretical framework posits supporting the planning instrument for territorial planning. It is potentially renewed, considering the process-oriented outlook of its parent stock. It practically exhumes some of the fading planning principles on the grounds of expedience. The new instruments are doubly faced with the weakness identified for existing instruments which is not disconnected with funding mechanisms and participatory problems. This underlines the need for government commitment, enhanced legal status of the instruments and stakeholders to enable them monitor funding and plan implementation through interactive participation.

10.1.5 Elements of Neo-mercantile Planning Theory

Neo-mercantile planning adopts integrated planning approach to address spatial planning at all administrative levels and to produce different framework instruments. It works in principle with a spatial model that defines activity belts and applies cumulatively across administrative boundaries to facilitate infrastructural and land-use development for spatial regional integration. In this sub-section, the

Table 10.2 Matrix of theoretical framework for spatial and statutory planning in Africa

	Objectives	Options	Criteria
Theoretical framework	The concept of urban environment	Systems approach	Study of system of base, deep structure, superficial structure
		Spatial approach	Delineation of urban core, urban outer ring (fringe area), urban outer ring (hinterland area)
		Urban structure approach	Concentric zone theory; sector theory; multi-sector theory
	Statutory planning	Provisional documents	Master plans; design concepts
		Operational documents	Planning schemes; Layout schemes; Action plans
		Regulatory documents	Normative space standards
		Statutory provisions	Neo-African development ideology Urban development strategy Spatial design models
		Standardization	Policy interpretation Growth visions
	Spatial planning	Urban form	Land use densities, patterns, functions
		Urban growth management	UGB, UDB, Greenbelt, Urban service limit, urban edge
		Use of space (Land use control)	Planning rationality; Market force
		Form-based planning concept	Spatial determinism; Spatial equilibrium; Urban growth management; Land use systems; 3-dimensional planning Functional flow analysis in space; Multifunctional space analysis (time budget); Asset-based analysis; Urban design standards
	Nature of planning	Developmental	Project (facility) plans; economic plans; assets utility plans
		Visionary	Long-term objectives; city vision statement; mind set; outlook issues
		Territorial	Long-term structure plans; detailed plans; planned unit developments (PUDs)
	Purpose of planning	Economic	GDP, Productivity; employment; use of resources; infrastructure planning, etc.
		Cultural	Conservation of heritage issues; cityscape concerns; civic identity concerns, etc.
		Health	Urban sanitation measures; urban quality control, etc.
		Form	Urbanity standards, etc.

(continued)

Table 10.2 (continued)

	Objectives	Options	Criteria
	Planning instruments	Planning initiative	Urban planning and local planning approach, etc.
		Planning perspective	Spatial equilibrium approach (SEA)
		Planning framework	Design concepts
		Thinking instrument	Neo-mercantilism; Neo-African planning paradigm
		Urban modelling	Merchant cities models
		Infrastructure development approval	Validation workshops (technical)
	Participatory process	Consultative	Professionals are under no obligation to take on board people's views
		Interactive	Groups take control over local decisions and determine how available resources are used
		Functional	Participation is seen by external agencies as a means to achieve project goals, especially reduced costs
		Passive	People participate by being told what has been decided or has already happened
		Self-mobilization	People participate by taking initiatives independently of external institutions to change systems
		Participation for material incentives	People participate by contributing resources, for example labour, in return for material incentives
		Logistics (pragmatic)	Mainstreaming monitoring; Participatory development; Feedback mechanisms; Intensive participation
	Planning methodology	Classic	Survey-analysis-design approach
		Rational	Participatory approach (for project planning)
		Neoclassic	New-master planning approach
		Standardized (normative)	Spatio-Metric planning approach
		Logistics (accreditation and technical forums)	Standard public sector planning departments; Researchers—consultants—policy makers forum
	Urbanism	New urbanism	Smart growth
		Sustainable urbanism	Design-oriented approach to planned development: relative density
		Creative urbanism	Culture-based urban design

(continued)

Table 10.2 (continued)

	Objectives	Options	Criteria
	Planning knowledge	Formal expertise	Professional practice; scientific database; use of planning theories and concepts, Urban design theory; etc.
		Informal expertise	Stakeholders forum; town-hall meetings, opinion poll, etc.
	Plan evaluation technique	Planning as control of the future	Non-implementation of plan
		Process of decision making	Decision making methodology; monitoring activities
		Intermediate technique	Plan alteration irrespective of implementation
	Regional integration	Spatial integration	Planning based on regional classifications: urban region, functional region, physical formal region, economic formal region, planning region, regional master plans, etc.
		Space economy	National urban development strategies (NUDS)
		Regional connectivity	Regional road network; functional flow-chart
		Economic integration	Economic reforms; political reforms, etc.
		Regional plans/budgeting	Urban market regions; Resource Management plans; Borderless worldwide cities concept
		Visioning process	Growth visions—spatial model—road maps
		Vertical planning	Intervention plan
	Cross-cutting issues	Heritage of city development in Africa	Civic identity
		Urbanization	Urbanization ideology; urban policy
		Urban growth	Spatial growth issues: qualitative, quantitative, structural growth, urban change, etc.
		Informal sector	Modern informal sector model
		Unlock resistance to planning through engagement	Cooperative governance
		Planning toolkit	Visual (graphic) presentation; spatial design concepts

Source Own construction (2013)

10.1 Neo-African Spatial Development Theory

levels of integrated planning, the structure of integrated planning and its instruments will be discussed. An integrated planning urban model and the alignment of integrated planning networks in a national grid will be proposed.

10.1.5.1 Neo-mercantile Planning Initiative

A neo-mercantile planning initiative is an integrated territorial planning concept. *Territorial* refers to *geographic space* defined by environmental conditions or political delineations or statutory provisions or any authentic regionalization. Integrated planning applies to these space delineations according to spatial coverage descending from the apex thus: Trans-national, national, provincial and municipal, city, and village level. These levels of integrated planning are guided by a singular continental mindset and outlook that is interpreted at all levels of spatial integration. Integrated planning provides for funding mechanisms and intervention plans which make them a structural part of economic development planning and budgeting at different levels of administration.

Integrated territorial planning is a form-based planning operation aimed at achieving time-efficient cities. The form-based element of the planning concept is considered at two levels: first within the urban core where it addresses urbanity through the interplay of land-use systems, distribution and patterns in space; and second at the urban regional level where it identifies activity belts based on the spatial definition of urban environment. Normal land-use planning processes reapplied on this spatial framework of integrated planning procedures to design time-efficient cities. Time-efficient cities underpin productivity which is initially factored on trade functions. This borders on a market space economy which could be introverted or extroverted, and either way is subject to innovations in the activity belts and the stages of development of the cities.

The integrated planning concept anticipates cities that undergo six stages of development in spatio-physical terms. Support for the stages of development draws from the antecedence of city development in Africa. Hence the stages include: unbuilt-up transient trade locations, built-up transient trade locations, maturity of trade nodes, suburban industrialization, development of industrial satellites and development of polycentric urban regions. Each stage confirms persistent positive appraisals of time efficiency in business transactions, irrespective of the productivity base as argued in export-base theory. Time efficiency relates to functioning without waste of time. Otherwise dissemination sets in as it had been the case with many traditional cities in the colonial period when the economic geography was rewritten. From stage three—maturity of trade node—they start to attract industrial and administrative functions.

The urbanity of core areas of cities is modelled according to the compact city concept in integrated planning initiatives. Because cities have to maximize the use of time they are modelled with the principles of compaction and the multifunctional use of space. The use of growth management instruments is inevitable and further identifies the planning initiative.

An integrated planning initiative is a political instrument. It precedes every political administration and provides political manifestos spatial model and road map for implementation. The roll-out model guides the post-roll-out models. Except for the roll-out initiative which defines the shared vision and long-term objective spanning thirty years, the post roll-out initiatives will last for the life span of the administration that initiates it. The preparation of integrated planning frameworks is provided as a central element of a transition programme for installing a new political administration. New administrations will remain provisional until the frameworks have been prepared and approved. The approval procedure will be interactively approved by a committee comprising executive members of the new administration, representatives of allied professional bodies and registration councils, organized civil societies, judiciary and invited resource persons. The committee will be chaired by a registered town planner appointed by the chief executive-elect and renowned for his contribution in integrated planning procedures.

The procedures for integrated planning require legal reforms directed at enhancing the powers and authority of the initiative. The reforms will deal with issues related to: mandatory measures regarding integrated planning, procedure of approval and use of prepared frameworks, time-frames for approvals, constitution of the committee, role of the planning machinery, organized civil society, power distribution in relation to economic development planning and budgeting, reviews, monitoring, etc. Of utmost importance is authority, because existing initiatives are found to lack authority. Therefore the provisions on the use of integrated planning frameworks are expected to be ruggedly binding, including provisions to legally challenge defined forms of side-lining. These reforms will be contained in a bill prepared by a committee of allied professional bodies and registration councils in the building industry led by a town planning registration council or professional body.

There are four categories of integrated planning initiatives, of which the first three are on the regional scale while the last is on the urban scale. The three integrated planning initiatives on the regional scale are: national, provincial, municipal and integrated regional planning, and the fourth category for cities is the integrated urban planning typology. The application of these integrated planning options requires two operations: first regionalization that is sub-divisions into market regions and the second grouping of cities happens in terms of size and function in the regional network. There are four categories of cities by function and they are: trans-national or ICT cities that coordinate information production functions for international trade relations, national cites or growth centres that coordinate service functions for internal trade relations and networking, provincial cities or agropolitan cities that coordinate manufacturing functions for economic growth and productivity, and municipal cities or agrovilles, that coordinate hamlets' primary functions for land resource management. The division of functions is not discrete but directional.

The different classes of integrated planning are aligned with three instruments, viz. the delineation of activity belts, infrastructure grids and city networks. Details of this alignment are discussed under framework instruments, which incidentally correlate with the dimensions of integrated planning initiatives. The dimensions of integrated planning are a visioning process of three stages of activities: visioning, spatial modelling and road-map construction. Spatial modelling has the components of urban space plan (core area), activity-belt plan (urban environment), and concept plan (urban region). The road-map construction components include sectoral master plans (projects) and land-use master plans (land resources). The structure of each plan is customized but they all adopt a management approach, which is discussed in the next section.

10.1.5.2 Neo-mercantile Planning (Approach) Perspective

The innovative drive of neo-mercantile planning favours a management approach which is encapsulated in a Spatial Integration Planning Approach (SIPA). The mission ahead of SIPA is to provide the spatial bases for resource management in space for neo-mercantile activities. To this end SIPA is committed at the outset to support trading in natural resources for economic growth, thus it combines economic and spatial planning to model time-efficient cities which will serve as crucibles for growth. In other words, SIPA essentially engages in assets utility planning in space to enhance distributive networks. It introduces time-budgets in spatial planning to address the demand-side and supply-side approach to regional growth as discussed under export-base theory and the basic strategy is to reduce the time traders (sellers and consumers) spend locating goods and services.

As a management tool SIPA functionally primarily engages in planning, but it does other functions, including coordinating, organizing and controlling. The overarching planning function is contained in the SIPA process which comprises a four-stage system of activities represented in Table 10.3. The SIPA process, classic planning process, neo-mercantile planning methodology and levels of participation in planning activities are related in Fig. 10.1. The purpose is to identify the relative

Table 10.3 Activity system of the SIPA process

Stages	Primary activities
Conceptualization	• Mindset and outlook issues • Growth vision • Long-term objectives
Analysis	• Spatial metric analysis • Remote sensing • Land use analysis • Environmental condition analysis • Asset-based analysis
Synthesis (Integration)	• Infrastructural planning • Spatial modelling • Planned unit development models
Implementation	• Facility (project) planning • Funding and budgeting • Interactive forum • Monitoring

Source Own construction (2013)

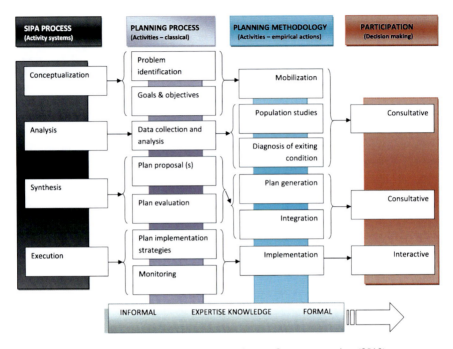

Fig. 10.1 Conceptualization of the SIPA process. *Source* Own construction (2013)

positions of activities in the many scenarios encountered in the process of delivering neo-mercantile planning initiatives. A characteristic feature of activity relationships is the positioning of sectoral (project) planning below the plan integration stage. Hitherto, in existing perspectives, sectoral (project) planning precedes integration. This was found to be responsible for the poor performance of integration processes.

The elements of participation in SIPA process are elaborated in Table 10.4 below. Who should be involved in the participatory process is conceived to include first, 'targets of change', that is the people at whom the intervention is aimed or whom it is intended to benefit; in some instances they are categorized as 'insiders'; and second, 'agents of change', referring to the people who make or influence policy or public opinion, also categorized as 'outsiders', and this includes external development partners. Participation is considered to be foremost at the point of visioning, modelling and funding initiatives and the methodology anticipates participation during the mobilization, planning studies and design and implementation stages of planning. The provisions made for participation are meant to complement the high-tech planning perspective which relies on but is not limited to remote sensing, Landsat and GIS data on land cover and spatial metric data processing. Caution was exercised to ensure that the creative input is protected and not traded off with political analysis that goes with project planning and funding mechanisms as it was found to be the case with existing planning instruments—rather participation is programmed to provide effective vanguard services against misappropriation of project funds.

10.1 Neo-African Spatial Development Theory

Table 10.4 Matrix of participation in SIPA process

Who (space, actors & knowledge)			Where (space/planning initiative)				When (planning process/methodology)			
Categories	Stakeholders		Planning initiative	Participation in planning (motivations and determinants)			1	2	3	4
				Visioning	Creativity (modelling)	Funding	Mobilization	Planning studies	Design	Implementation
Targets of change (Insiders)	Primary stakeholders - Upper class (elites) - Upper-middle class (professionals) - Lower-middle class - Low class - Proletariats (absolutely poor) Secondary stakeholders - Union leaders(informal sector) - Professional bodies		Territorial	Formal ≥ informal expertise knowledge	Formal ≥ informal expertise knowledge	Informal expertise knowledge	Interactive	Consultative	Consultative	Interactive
			National	Formal ≥ informal expertise knowledge	Formal ≥ informal expertise knowledge	Informal expertise knowledge	Interactive	Consultative	Consultative	Interactive
			Province	Formal ≥ informal expertise knowledge	Formal > informal expertise knowledge	Formal ≤ informal expertise knowledge	Functional	Consultative	Consultative	Interactive
Agents of change (Outsiders)	- Public sector administrators - Private sector - Freelance activists (opinion leaders) - Consultants (indigenous & foreign) - Motivators (Politicians & opinion leaders) - External development agents		Municipal	Formal > informal expertise knowledge	Formal expertise knowledge	Formal < informal expertise knowledge	Consultative	Consultative	Consultative	Interactive
			City	Formal = informal expertise knowledge	Formal expertise knowledge	Informal expertise knowledge inclusively	Consultative	Consultative	Consultative	Interactive

Source Own construction (2013)

In the next section, integration planning working instruments will be discussed. As mentioned earlier, the instruments are used for the visioning process for different scales of spatial systems and administrative boundaries. All instruments are concerned with creative spatial proposals in space either broad-based or detailed.

10.1.5.3 Neo-mercantile Planning Framework

The foremost neo-mercantile planning frameworks are contained in spatial integration plans (SIPs). The SIPs are supported with a wide range of thematic integration plans (TIPs). While the SIPs are broad-based plans, the TIPs are detailed plans and they relate complementarily. They summarily provide codes of conduct loosely referred to as guidelines for the use of space. The frameworks are managed through consultancy services or in-house professional expertise or both. There are four SIPs as represented in Table 10.5.

The SIPs take precedence in descending order as presented in Table 10.5. The next set of framework instruments are the TIPs. These categories of instruments and their major activity focus are summarized in Table 10.6.

SIP is the central planning instrument at all levels of administration. It combines economic and spatial planning and incorporates an investment plan. All other forms of development planning including budgeting at all levels of administration will be done in the context of SIP guidelines and provisions. In other words, SIP is the visioning instrument. The flow chart of the instruments is represented in Fig. 10.2.

Table 10.5 Categories of SIPs instruments

Administrative level	Context (region)	Activity overview	Major line agencies
National	National spatial development plans	Focuses on the national grid for infrastructural development. Provides guidelines for the development of the trans-national and national market regions. Distribute national development projects in different categories of activity belts. Accounts for development plan of the core area of ICT cities	National Planning Commissions, Federal Ministries, etc.
	Spatial development plans for Trans-national market regions	Focuses on shared vision, policy development and trade relations interpretation to guide planning activities. Coordinates communication network and distributes trans-national projects in the core areas of ICT cities	AU and other regional organizations egg, SADC, ECOWAS, etc.
	Spatial development plans for national market regions	Focuses on political manifestos (visions), trade and economic development and policy interpretation. Accounts for the development plan of the core area and urban environment of Growth centres (national cities)	National Planning Commissions, Federal Ministries, etc.
Provincial	Spatial development plans for provincial market regions	Focuses on the provision of provincial grid for infrastructural development. Provision of integrated plan for modelling the urban regions via, facilities distribution and regional connectivity. Provision of guidelines for development priorities in activity belts. Accounts for the development plan of the core area and the urban environment of Agropolitan cities (provincial cities)	Provincial departments of planning and sectoral departments
Municipal	Spatial development plans for municipal market regions	Focuses on functional flow and land use budget in the activity belts as well as the provision of development plan of core areas of Agrovilles (municipal cities) within the guideline of the provincial plans	Local Planning Authorities, etc.

Source Own construction (2013)

10.1 Neo-African Spatial Development Theory

Table 10.6 Categories of TIPs instruments

Thematic Instruments	Spatial system	Activity overview	Major line agencies
Urban core plans	Core area	Urban design. Focuses on growth boundaries, urban from. Time-budget principles for land use planning and transportation networks. Urbanity. Village design	City development authorities, Town unions
Activity-belt plans	Urban environment	Modelling of the urban form through the spatial distribution of activities in the core area, inner ring, and outer ring of the urban environment. Industrial sites	Municipalities, city development authorities
Concept plans	Urban region	Structure plan for urban region integration	Municipalities.
Land-use master plans	Land resources	Land resource planning in the outer ring for agriculture, mining, conservation, etc.	Municipalities
Sectoral master plans	Projects	Project planning	Specialized agencies

Source Own construction (2013)

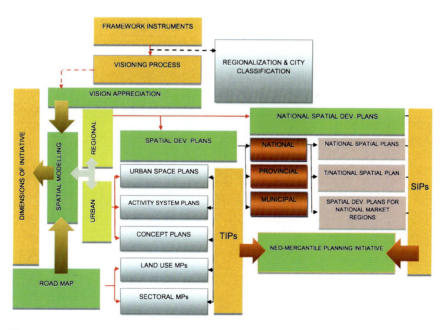

Fig. 10.2 Flow chart of neo-mercantile planning framework. *Source* Own construction (2013)

The SIP is contained in two reports—Diagnostic and Integration reports. The diagnostic report establishes the status quo vis-à-vis the system of base, activity space relationships and the superficial structure of the subject under investigation. In this way environmental factors including cultural and value systems, spatial determinism and spatial equilibrium are determined using a spatial metrics tactic of remote sensing and GIS data-collection methodologies. The data collected is used for resource analysis as well as form-based analysis of built-up areas. Resource analysis is stratified into natural (renewable and non-renewable) and human resources (personal and community assets and capacities) while form-based analysis focuses on urban growth management issues related to land use and population distribution, densities, patterns and dynamics, growth management instrument and connectivity. The expected output which the report contributes to SIPA process is the identification of priority problems, the establishment of long-term spatial integration objectives and the simulation of growth visions with measurable indicators.

The SIP integration report performs four functions: first, it generates action plans, second it identifies tasks for the action plans, third it configures a spatial model for distributing the tasks, and fourth, it proposes an investment plan. Each of the four functions is a composite set of activity performed through participatory processes, controlled, however, by formal expertise knowledge in spatial planning. The action plan takes care of centrally considered sectoral planning activities.

The modelling is for the adjustment of the basic scenario and alternative proposals will be evaluated using investment criteria. The proposals will be in the categories of heavy investment, light investment and heavy/light investment proposals. The evaluation will be done with a predetermined template and the chosen concept further elaborated, leading up to strategic planning for funding and implementation. To this end costing and institutional frameworks are determined and an investment plan proposed.

The frameworks have multiple relationships and alignments as the space they act on overlap and so do the spatial systems vary in categories. The resultant spatial network is captured in the next section by way of grid diagrams.

10.2 Spatial Integration Network

The SIPs and TIPs act on five levels of the administrative framework, six categories of spatial systems, and two conceptions of three activity belts. The taxonomy of cities identifies four categories of cities, excluding rural communities and hamlets, and each class of cities is responsible to a level of government as mentioned earlier and represented in Table 10.7.

Each spatial unit has activity belts in micro- and macro-dimensions. The micro-dimension relates to its structure as a spatial system with a core area, inner ring area and outer ring area, although the spatio-physical extent of the three spatial space varies depending on the spatial unit as illustrated in Fig. 10.3.

10.2 Spatial Integration Network

Table 10.7 Categories of spatial systems

Administrative level	Spatial systems
National	ITC cities
	Growth centres
Provinces	Agropolitan cities
Municipals	Agrovilles
	Rural communities
	Hamlets

Source Own construction (2013)

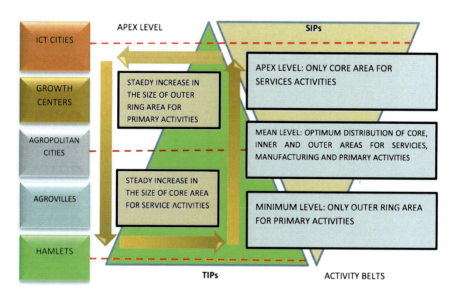

Fig. 10.3 Pyramids of framework instruments related to the categories of cities. *Source* Own construction (2013)

Agropolitan cities represent the optimal situation where the spatial spaces even out, while at the lowest grade of spatial unit, that is, hamlets, the outer ring area dominates and at the top, that is ICT cities, the service area dominates conceptually. There is a continuum in the dynamics of the spatial space for each spatial unit. The spatial spaces are referred to as activity belts as they are assigned service, manufacturing and primary functions respectively. The SIPs and TIPs act inversely on the micro-activity belts. While TIPs are maximally used at the lowest level and diminish towards the apex, SIPs are maximally used at the apex and diminish towards the lowest level.

The macro-activity belt is conceived in the context of regional spatial integration to determine the catchment area of the different classes of spatial systems and by implication administrative levels. This is illustrated in Fig. 10.4.

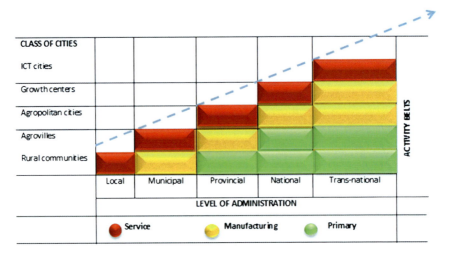

Fig. 10.4 Composition of the work environment of SIPs and TIPs

For instance, the catchment area of ICT cities comprises the Growth centres and Agropolitan cities located within the service belt and the Agrovilles and rural communities located within the primary belt. Note that rural communities and hamlets do not have catchment areas. The activities of national government in the catchment area of ITC cities are guided by the activity belt sub-division. For example, if the national government wants to locate a national abattoir, an ITC city is identified for that purpose, using mainly political analysis and within the manufacturing belt of the identified ITC city's catchment area a suitable site is identified using political, economic and spatial analysis in ascending order.

The relationship between the spatial systems is hierarchical. Hence there are five vertical surfaces; each identifies a grid of connectivity (transport and communication routes) that networks a class of spatial systems as shown in Fig. 10.5. The dominance of use of TIPs and SIPs for planning activities as in the micro-classification of activity belts alternates within the macro-activity belts.

The relationship between the five grids is pronounced when they are collapsed into a single surface. This represents the horizontal relationships represented in Fig. 10.6. The higher the class of spatial unit the more the spatial systems in lower class it has to download in the management of economic growth. Several of each class of spatial unit is usually identified and they compete and cooperate among themselves interactively and complementarily in infrastructural development and business ventures. The spatial systems serve as intervening opportunities factored on time-efficiency principles and of course other productivity imperatives. The functional relationships facilitate the identification of development corridors, growth triangles and ultimately the delineation of market regions. The market region matrix will inform economic development plans at all administrative levels. For administrative convenience, market regions and administrative boundaries may

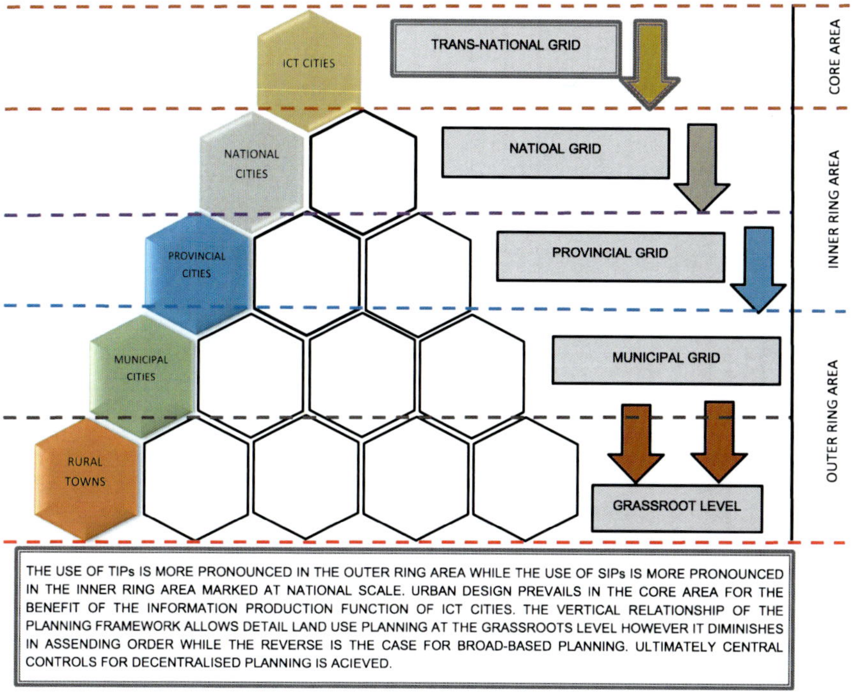

10.2 Spatial Integration Network

Fig. 10.5 Vertical relationships in spatial integration network. *Source* Own construction (2013)

not overlap to the effect that a combination of provinces could make-up a market region. The market regions will be networked at the trans-national level into 'trade-basins', spatio-physically expressed in trans-megapolitan development.

As illustrated in Fig. 10.7, six trade-basins are discerned, based on regional cooperation, particularly prior colonization and current regionalization tendencies. The forest area coastal trade-basin, covering ECOWAS countries plus Cameroon, Gabon, Equatorial Guinea; the Western Sudan trade-basin covering Mauritania, Mali, Niger, Chad; the Central Africa trade-basin covering the Central African Republic, Congo, Zambia, Zimbabwe, Zaire, Rwanda, Malawi, Uganda, Burundi; the South-east Coastal trade-basin covering South Africa, Mozambique, Tanzania, Kenya, Swaziland, Lesotho; the Atlantic Coastal trade-basin covering Angola, Namibia, Botswana; and the Horn of Africa Coastal trade-basin covering Sudan, Ethiopia, Somalia and Djibouti. Except for the Central African trade-basin, the rest are all potential trade destinations for international trade.

The four coastal trade-basins will be deployed to service Euro-American trade relations while the Western Sudan trade-basin will address trade with the Arabs in the middle-east via North Africa. The Central African trade-basin will coordinate internal trade among the six trade-basins. Given this framework, countries with potentials to produce megacities that will serve as principal trade nodes are Nigeria,

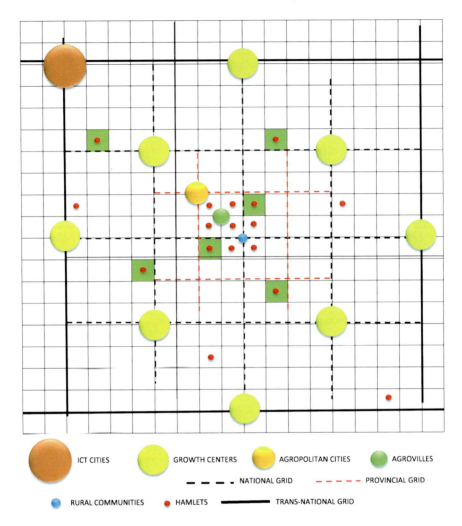

Fig. 10.6 Horizontal relationships in spatial integration network. *Source* Own construction (2013)

Zaire, South Africa, Kenya, Sudan, Angola and Niger. For spatial reasons, Senegal will be encouraged to serve as principal international trade node country, although Senegal and Niger in terms of population do not have the potential to generate megacities. Overall, the proposal views trade relations from a spatio-physical perspective and in the process provide a platform for considering new regionalism for Africa.

The next section builds on the argument of form and function to generate a design model for urban region development. The design is aligned with the spatial growth processes of polycentricism in space economy. Therefore the design adopts

10.2 Spatial Integration Network

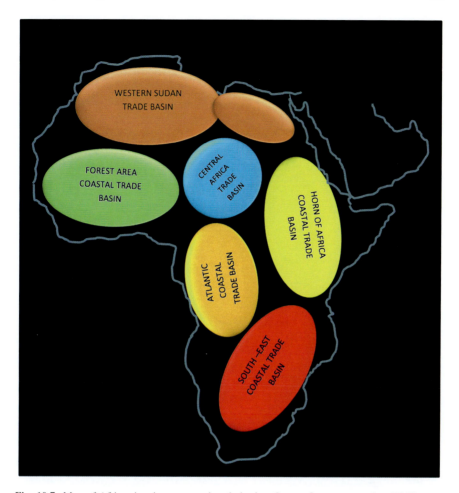

Fig. 10.7 Map of Africa showing proposed trade basins. *Source* Own construction (2013)

activity belt strategies to aid spatial growth processes. This is strategic for the integration of vertical and horizontal relationships in the market regions with extended metropolitan regional and growth triangle models.

10.3 Urban Spatial Model

The basic outline of the urban spatial model is illustrated in Fig. 10.8. The design approach is based on the spatial definition of the urban environment. The three sections of the urban environment are regarded as activity belts. The core area is allocated service (marketing) functions (showrooms and shopping malls), the inner

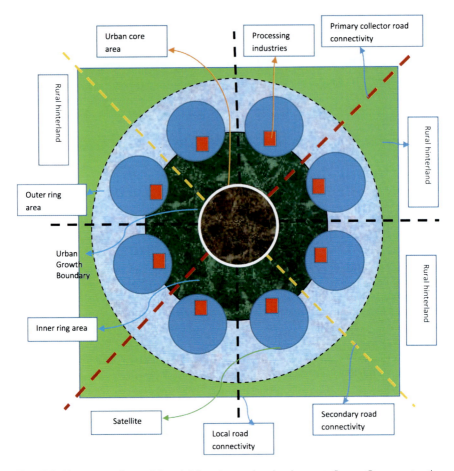

Fig. 10.8 Neo-mercantile spatial model for urban region development. *Source* Own construction (2013)

ring area performs manufacturing (extraction and processing) functions (minerals and agro-based industries), and the outer ring area, including the rural hinterland, performs primary economic functions (farm plantations and mining sites). The core area is hemmed-in and its growth managed with an urban growth boundary, which implies time budgeting in land-use planning in the urban region. In other words, the spatial relationship between Brownfield and Greenfield development is timorously managed.

The design defines the space for urban economy to extend beyond the core area of the urban environment and extends into the countryside. The approach to use the spatial space is strategic, to encourage polycentric urban form. The polycentric urban form is built on functional flow principles for the urbanizing city. It encourages the distribution of land use that facilitates functionality and region-based urbanization. With the distribution of processing industries, provision is made for the development of new satellites in the growth process. The satellites are linked with regional road networks and inter-connectivity pronounces the extended metropolitan model.

10.4 Measure of Time-Efficient Coefficient for the Classification of Settlements

In early twentieth century functional classification models using dominant functions became commonplace. Some of these models, which were based essentially on changes in socio-economic indicators, included "Harris's Functional Urban Classification" (Harris 1943); "Nelson's Multifunctional Classification" (Nelson 1955); "Alexanderson's Method" (Alexanderson 1956); "Webb's Analysis of Minnesota Towns" (Webb 1959); "The Duncan and Reiss Classification" (Duncan and Reiss 1956); and "Forstall's Classification of American Cities" (Forstall 1970). In a different representation, Bechtel (1973) summarized new contributions thus:

> Duncan (1960) represents the functional or economic viewpoint from which to classify cities. Cox and Zannaras (1970) represent an attempt to classify cities by popular views of the geographical region to which they belong. Maloney (1967) represents a purely factorial approach and has two classifications if one counts those cities loading both high and low on his factors. Forstall (1970) represents the City Manager's Yearbook method of classification and actually has three methods of classification. Nelson (1955) classifies cities according to a service criterion.

Using factor analysis and later cluster analysis, studies were conducted with these models in the late twentieth century, especially Nelson's model. Since the turn of the century the inclusion of spatial indicators has identified the latest contributions, although this trend is still in the preliminary stages and results are tentative, requiring further analysis. Drewes and Van Aswegen (2011: 15–25) contributed a related study on the vitality of urban centres in which spatial indicators were used to classify urban centres in Northern Cape Province, South Africa. Drewes and Van Aswegen's study, however, was primarily based on the industrial theory that explains the spatial incidence of urban areas.

Under neo-mercantile planning the ranking of cities or settlements within different categories of spatial systems will be done by calculating the time-efficient coefficient of the settlements. Two approaches can be used to determine the coefficient (herein after referred to as TE-coefficient)—statistical and theoretical approaches. Expert opinion from the Mathematics Department and the Statistics Consultancy Service Unit at the NWU indicates that the theoretical approach is a complicated process requiring further research. The statistical approach is basically the objective determination of weighted means calculated from variables that are put on a standard scale and then weighted by means of a statistical process.

The variables as indicated earlier for measuring time-efficiency of spatial systems are potentially on different scales. Time-efficiency is dependent on pull-factors or the attractiveness of spatial systems as business location. The pull-factors have three dimensions, namely social, economic, and spatial. The social dimension of the variables includes innovation, standard of living, age, sex, occupation, administrative functions, etc. The variables for economic dimension include entrepreneurship, transport costs, modal splits, traffic delays, opportunity costs, induced demand, trade relations, etc. For the spatial dimension the variables include location, size of catchment area, accessibility, proximity, Weigh-bill transfers, trip assignment, urban form, functional flow, etc.

The statistical approach of calculating time-efficient coefficient is in three phases, identified thus:

Phase one—identification of k independent variables: $X_1, X_2, X_3, X_4, \ldots X_k$.

Phase two—put the variables in standard scale using the formula:

$$Y_i = \frac{X_i - \overline{X}_i}{S_i}$$

where:
Y_i the ith standardized variable, $i = 1, \ldots, k$
X_{i^1} the ith variable, $i = 1, \ldots, k$ $\overline{X}_{i^1}$ = Mean value of X_i
S_i Standard deviation of X_i

Phase three—a principal component analysis on the $Y'_i s$, from which the first component score can be obtained and is mathematically represented as:

$$T_E = w_1 y_1 + w_2 y_2 + w_3 y_3 + w_4 y_4 + \cdots + w_k y_k.$$

where:
T_E TE-coefficient (weighted mean of the standardized variables), and
$w_1, w_2, \ldots, w_k$ are the weights such that $w_1 + w_2 + \cdots + w_k = 100\ \%$

Such a coefficient will only make sense if the first principal component explains a substantial percentage of the total variance of the standardized variables.

10.5 Conclusions

The neo-mercantile planning theory is committed to handling humanistic interventions responsible for urban change given the projected population growth and urbanization and urban growth phenomenon anticipated in the African region in the new millennium. It does so without prejudice to environmental determinism because human-induced land use changes are considered the prime agents of global environmental changes (Ramachandra et al. 2012). Thus neo-mercantile planning theory derives from spatial metrics environment to address spatial equilibrium in tandem with spatial determinism in economics.

Neo-mercantile planning strategically elevates spatial planning from its very low pedestal to an apex position where it exerts an overarching influence overall other

10.5 Conclusions

forms of development planning. Within its process, sectoral and project planning processes relocate to merge with plan implementation procedures which are subject to the provisions of spatial integration. Budgeting is also strategically subsumed into the planning system with an investment plan and funding strategies which forms part of the neo-mercantile planning framework.

A major contribution of neo-mercantile planning is its spatial model for urban region development. The model facilitates activity-belt strategies which redistribute activities within the urban landscape. This has the potential to translate growth visions of urban Africa into space. The instruments for securing the model are grouped under spatial integration planning (SIPs) and TIPs instruments. Three levels of planning and six categories of spatial systems will be integrated in line with the standards of extended metropolitan regional and growth triangle models for isolated market regions. Hence the new planning paradigm is thought to provide spatio-physical bases for integration. It demonstrates that planning rationality can still prevail in a market force context.

The new planning paradigm is mindful of the sentiments of participation in the planning process. Therefore provision is made not because the sentiments as expressed are considered correct in the context of neo-mercantile planning but because it is better thought of as a missing aspect of plan implementation. To this end, participatory provision in the new paradigm empowers monitoring of plan implementation procedures. It advocates a consultative outlook during creative processes which requires formal expertise knowledge although politicians will be engaged.

Neo-mercantile planning is linked with politics. It provides political manifestoes the spatial model and road map for their translation into space. This is why it is mainstreamed in the inception of a new political administration through its framework instruments. Therefore its adoption requires very strong political will acceded to by the committee of nations in the AU. Although the next section articulates the contribution made in terms of new knowledge, it is largely concerned with the domestication of the proposed planning theory in the African context. A domestication programme is subsequently proposed.

References

Bechtel RB (1973) Types of cities: the subnational urban environment and some design implications. In: EDRA 4. Fourth international EDRA conference, vol 1, pp 150–160

Bibangambah JR (1992) Macro-level constraints and the growth of the informal sector in Uganda. In: Becker J, PedersenP (eds) The rural-urban interface in Africa, Uppsala. The Scandinavian Institute for African Studies

Christaller W (1933, 1966) Central places in Southern Germany. (trans: Baskin CW). Prentice-Hall, Englewood Cliffs

Drewes JE, van Aswegen M (2011) Determining the vitality of urban centres. In: Brebbia CA (ed) The sustainable world. WIT Press, United Kingdom, pp 15–25

Duncan OD, Reiss AJ (1956) Social characteristics of urban and rural communities. Wiley, New York

Edwardo LE (1990) Community design and the culture of cities. Cambridge University Press, Cambridge, 356p

Forstall RL (1970) A new social and economic grouping of cities. The Municipal YearBook, pp 102–159

Friedmann J (1972) A general theory of polarized development. In: Hansen NM (ed) Growth centres in regional economic development. Free Press, New York

Harris CD (1943) A functional classification of cities in the United States. Geogr Rev 33:86–99p

Herold M, Couclelis H, Clarke KC (2005) The role of spatial metrics in the analysis and modelling of urban land use change. Comput Environ Urban Syst 29(2005):369–399

Landman K (2004) Gated communities in South Africa: the challenge for spatial planning and land use management. Town Plann Rev 75(2):163p

Nelson HJ (1955) A service classification of American cities. Econ Geogr 31:189–210

North DC (1956) Exports and regional economic growth: a reply. J Polit Econ 64(2):165–168

Poulicos P, Chrysoulakis N, Kochilakis G (2012) Spatial metrics for Greek cities using land cover information from the Urban Atlas. In: Gensel J et al (eds) Proceedings of the AGILE'2012 international conference on geographic information science, Avignon, 24–27 April 2012

Ramachandra TV, Bharath HA, Sreekantha S (2012) Spatial metrics based landscape structure and dynamics assessment for an emerging Indian Megalopolis. Int J Adv Res Artif Intell 1(1):48–57

Rostow WW (1977) Regional change in the fifth Kondratieff upswing. In: Perry DC, Watkins AJ (eds) The rise of the Sunbelt cities. Sage, Beverly Hills

Tiebout CM (1956) Exports and regional economic growth. J Polit Econ 64(2):160–164

Webb JW (1959) Basic concepts in the analysis of small urban centres of Minnesota. Ann Assoc Am Geogr 49(1):55–72p

Weber A (1929) Theory of the location of industries. University of Chicago Press, Chicago

Part IV
Contribution to New Knowledge, Application and Approaches

Chapter 11
New Knowledge in Planning (in Africa)

Abstract The visioning process for domesticating neo-mercantile planning theory in Africa identifies priority transboundary problems to include urban spatio-physical expansion, suburbanization, urban sprawl, extroverted urban economies and spatial inequalities. With the vision exposition focused at building African civilization as a culture of cities that are integrated across national boundaries, the short-term objectives of creating enabling environment leads the mobilization of concrete action to resolve the priority transboundary problems within 15 years. Hence priorities for action are identified alongside performance indicators. These actions are regrouped into ten typology of action. International funding is likely to threaten the implementation of proposed action cards (although the action cards are low investment ventures; therefore not really vulnerable) but a lot depends on the resolve of Africa and the Diaspora to chart her destiny.

Keyword Visioning · Civilization · Extroverted · Mobilization · Implementation · NEPAD · Funding

11.1 Visioning Process for Spatial Integration Network in Africa

Spatial regional integration has eluded Africa for a long time and thereby hindered its long-standing battle for poverty reduction. This research contributes a theoretically compelling solution and attempts to align the new paradigm with the status quo. In the 1960s, the Organization of African Unity (OAU) was established to provide an institutional framework for regional integration in Africa. As contained in an **AU** web page

> On 9th September, 1999, the Heads of State and Government of the Organization of African Unity issued a Declaration (the Sirte Declaration) calling for the establishment of an African Union (AU), with a view, inter alia, to accelerating the process of integration in

the continent to enable it play its rightful role in the global economy while addressing multifaceted social, economic and political problems compounded as they are by certain negative aspects of globalization.

With this declaration, the mindset for establishing the AU is not in doubt as well as the AU's integration agenda for building a united and strong Africa.

The AU inherited the challenge of reducing poverty and was responsible for OAU initiatives—amongst the initiatives is the New Partnership for Africa's Development (NEPAD) initiative. The NEPAD initiative was adopted as a programme of the AU at the 2001 Lusaka Summit and its objectives focused on poverty reduction, sustainable growth and development, halting marginalization of Africa in globalization process and beneficial integration into the global economy. These are fundamental concerns that addressed the precarious position of Africa in the world system which NEPAD strategies tend to side-line in favour of MDGs.

The MDGs are more or less cross-cutting elements favourably disposed to sectoral planning. Unlike the MDGs the objectives of NEPAD are central elements in territorial planning for regional integration. The synthesis of the two scenarios is seldom considered—rather rivalry is commonplace in which territorial planning in Africa is endangered by global forces. Nevertheless there is a need for a spatial framework within which sectoral planning for MDGs should be deliberated while fate is maintained simultaneously with territorial planning. But this mechanism depends on the appreciation of transboundary problems and the shared vision of developing the space economy in Africa.

11.1.1 Priority Problems

The transboundary problems considered were fewer than five headings, namely urban spatio-physical expansion, suburbanization, urban sprawl, extroverted urban economies and spatial inequalities. Also considered as priority problems are informalization and the so called rapid urbanization. The scenario in select African countries is outlined in Table 11.1 below.

11.1.2 Vision Exposition

Based on the distortions and declining urban productivity a diagnosed territorial planning vision for Africa is encapsulated in a Vision Statement stated below

> African civilization built on a culture of cities that is integrated across national boundaries and arranged in a hierarchical system of functional relations to maintain systemic relevance in global mercantilism, reinforce per capita productivity and create wealth for enhanced African economy.

Table 11.1 Transboundary problems in selected African countries

Countries	Transboundary problems
DRC	• Sprawling informal expansion of the urban system into suburban areas in reaction to the suburbanization of poverty that derives from extroverted dualistic urban economy.
Angola	• The enclave-style economy coupled with insecurity in the hinterland areas informs the proliferation of unproductive informal settlement development in the peri-urban areas, leading to uncontrolled spatial spread of the primate city of Luanda to absorb nearby towns.
Mali	• Informal physical expansion of Bamako into thinly populated outlying areas as extroverted urban economy drives the impoverished society into survivalist informal sector that disconnects from the husbandry of local resources.
Egypt	• Sprawl expansion is commonplace in the peri-urban areas leading to uneven spatial structures that support an extroverted urban economy.
Senegal	• The leapfrog model of urban sprawl coupled with suburbanization of population caused mainly by land market-driven functional rearrangement leads to the growth of informal settlements in peripheral areas.
Kenya	• Extroverted urban economy causing rapid growth of informal economy and corresponding development of informal settlements mostly at the fringe areas as poverty driven urbanization leads to urban expansion.
Nigeria	• Urban expansion mainly in the form of commercial ribbon streets and peripheral slum development programmed to promote trade in foreign goods.
South Africa	• Spatial fragmentation and incipient low-density spatial expansion leading to edge cities and continual mushrooming of informal settlements on the urban edge.
Tanzania	• Introverted but survivalist economy that encourages Greenfield development in the form of informal subcentres in peri-urban areas—as spatial form of Dar es Salaam increasingly assume ribbon-like leapfrog pattern of land use distribution.
Ethiopia	• Poorly developed urban economic base and the incidence of squatter settlements leading to slum formation especially in the peripheral area of Addis Ababa.

Source Own construction (2014)

11.1.3 Vision Objectives

The focus on mercantilism is considered an entry point for wealth creation that will eventually culminate in production. The transition from mercantilism to production is a long-term programme, the foundation of which is laid with the restructuring of space economy to make it more compliant with productivity. The vision objectives are therefore meant to support measures that will restructure the space economy.

11.1.3.1 Short-Term Policy Objectives (in Five Years)

The short-term policy objectives focus on creating an enabling environment at continental level for take-off. Therefore short-term policy objectives to be achieved within five years include

  i. To secure political will for engaging neo-mercantilism as development ideology for Africa,
 ii. To prepare and adopt the National Integrated Regional Development Act (NIRDA),
iii. To delineate the trade basins in tandem with intra-regional development corridors,
 iv. To secure complete regeneration of AMCHUD and set up country-based technical units responsible for transboundary regional integration,
  v. To enact Integrated Regional Development Act (IRDA) within five years,
 vi. To setup a functional research institute in the AU for monitoring spatial integration networks.

11.1.3.2 Medium-Term Policy Objectives (in Ten Years)

The medium-term policy objectives focus on preliminary engagements at national level for the mobilization of concrete action. Therefore medium-term policy objectives to be achieved within 10 years include

  i. To domesticate the National Integrated Regional Development Act (NIRDA) at the national level,
 ii. To finish the classification of cities,
iii. To finish the grading of infrastructure grids,
 iv. To prepare spatial integration plans (SIPs) and thematic integration plans (TIPs) at country level,
  v. To fully mobilize plan implementation process with roll-out plans at country levels.

11.1.3.3 Long-Term Policy Objectives (in 15 Years)

The long-term policy objectives which will be achieved within 15 years provide directional guidelines for resolving priority transboundary problems. Primary amongst the transboundary problems are issues related to dispersal patterns of urbanization which co-exist with the inequity of urban primacy, extroverted urban economy characterized by survivalist informal sector, declining urban productivity and per capita productivity, urban planning that is divorced from investment planning, etc. Therefore the long-term policy objectives include

11.1 Visioning Process for Spatial Integration Network in Africa

i. Maximize by 75 % the use of space in urban core area within ten years,
ii. Discourage and reduce to half (50 %) indiscriminate land-use development, beyond urban growth boundaries and the incidence of informal settlement development by year 2024,
iii. Reverse 75 % of dispersed land-use distribution in urban core areas and institutionalize the regeneration of slums within 15 years,
iv. Redistribute economic land-use function in space within 15 years to encourage local productivity,
v. Discourage at least 75 % of isolated urban hierarchy and enhance integrated spatial development within 15 years,
vi. Maximize the use of urban growth management instruments within ten years (to introduce a time-element in spatial development), and
vii. Encourage extroverted urbanization and secure 50 % compliance within 15 years.

11.2 Priority Actions

The anticipated territorial planning frameworks for delineated trade basins as enunciated in neo-mercantile development ideology elucidate spatial integration networks underpinned in neo-mercantile planning protocol. Accordingly two operations are imperative; first the classification (or grading) of cities using administrative criteria and time-efficient coefficients, and, second, the scaling (or grading) of infrastructure grids using administrative criteria. The neo-mercantile protocol upholding the city as the basis for regional integration led by urban policy orientation of equitable spatial polarization. The installation process adopts the spatial integration planning (SIPs) instrument characterized by the hierarchy of cities which are networked at two levels: infrastructure network and functional flow relationships. The infrastructure fabric that networks the cities is a combination of grids calibrated at administrative levels. The national grid, for instance, synchronizes with national cities. On the other hand, functional flow criteria drawn from the time-efficient evaluation.

11.2.1 Assessment of Current Actions

Current actions on integrated regional development are summarized in the NEPAD initiative (c.f.11.2.1.1). This initiative is limited to economic and political reforms and lacking in spatial reforms. It sought economic bases of integration as is the case with prevailing regional development theories (c.f. 6.6.2). This reflects in AU/NEPAD action plan 2010–2015 which is a compendium of sectoral projects mainly focused on infrastructural development. Spatial content sought through the

Resource-based African Industrial and Development Strategy (RAIDS) initiative engaged development corridors strategy and to a lesser extent spatial development initiative (SDI), both determined by market forces and not planning rationalities (c. f.4.6.6.2). The only spirited effort at spatial planning is contained in the NEPAD cities initiative which is more or less a UN-Habitat brain child focused on slum upgrading projects. Rather than pursue NEPAD's central objectives of regional integration, the NEPAD cities' initiative is geared towards the delivery of MDGs and pro-poor measures in urban development. Informality rules its outlook as a planning instrument and in practice it has no spatial bases to fulfil its theoretical role as contained in the visioning process.

The institutional arrangement reflects the redundancy of AMCHUD and the transition of UN-Habitat position from support to control (c.f.1.1). The reliance of the financial mechanism of the initiative on foreign direct investment (FDI) is decried. Moreover, legislative support for NEPAD initiative is not clear. These institutional issues practically paralyzed NEPAD initiative and rendered it redundant and moribund after more than one decade of existence.

11.2.1.1 NEPAD Initiative

The general impression in African development planning literature according to institutionalist content drivers is that NEPAD initiative is built on political and economic reforms that are geared towards regional integration for Africa. NEPAD advocates good governance, conflict resolution, the rule of law, macro-economic stability and curbing of corruption (Richard 2002). It seeks to provide a comprehensive framework on which 'Africa can collectively and effectively co-operate with its development partners' (Richard 2002). Thus the NEPAD framework incorporates the African ministerial conference on housing and urban development (AMCHUD) to manage urban growth; the NEPAD Business Group (NBG) to manage business enterprises; Ministers of public service to manage good governance; and the African Peer Review mechanism to share best practice among African countries. The literature seldom provides some invaluable insights that represent reality.

A little bit of insight into NEPAD shows it has gone by various names, including the African Renaissance (1996–2000), the Millennium Africa Recovery Plan (2000–July 2001) and the New African Initiative (July–October 2001) (Bond and Dor 2003). Remarkably NEPAD evolved in close contact with the G8 (Bond and Dor 2003: 24) and the Bretton Woods Institutions and international capital (Bond and Dor 2003: 24), under conditions of secrecy. NEPAD lost its original outlook and by 2001 it has become another form of structural adjustment in disguise (Bond 2002 and http://www.nepad.org). In fact, Bond and Dor (2003) view NEPAD as home-grown African neoliberalism. Nevertheless, in line with neoliberal values, NEPAD has since undergone official state and business endorsement processes.

11.2 Priority Actions

In its neoliberal outlook, NEPAD initiative lacks a concrete vision statement and growth indicators. It is incapacitated to play the role set out in its stated objective. For instance, the focus of NEPAD's attention on poverty eradication does not reconcile with the original objective of NEPAD which point towards the reversal of trends in the imperial economy of African states. The focus of NEPAD is instead hijacked to address MDGs, courtesy of imperialism as international administration. This trend has generated critical views about the potential of NEPAD to provide the required lead that will pull Africa out of the woods. The common impression is that the new ideology seems to disregard the epistemological foundation of African problems. Indeed Nabudere (2003) believes that dismantling the old economic order will not be achieved if we are to address the kind of issues that NEPAD tries to address and which it avoids to address.

The NBG is reported to be making contributions especially in South Africa, Lesotho, Nigeria and Kenya (Tawfik 2008: 66). Most probably they are coordinating business relations with foreign partners in line with the neoliberal outlook of NEPAD. Meanwhile, in the political realm three trends are embedded in the positive appraisal of democracy and good governance in Africa that tend to challenge the vision of engaging a development paradigm for the NEPAD initiative. First, the political institutions are consistently adjudged to be weak, thus requiring capacity building; second, increasing association of the political system with participatory process; and third, it seems the more African governments are responsive to imperial control the more they are adjudged to be democratic in the perception of the global north. These trends somehow adversely influence African political economy in such a way that the imperial economy bequeathed to Africa since colonization remains resilient. This is why the attention on eradicating poverty should be properly diagnosed and handled carefully. Its relevance in conceptualizing development paradigm for Africa is beneath the need to rework the imperial space economy in Africa.

The supportive activities of external assistance agencies that usually slide into control are some of the legitimate fears that critics express. The UN-Habitat was instrumental to the formation of AMCHUD in 2005 and ever since has practically determined development projects while AMCHUD religiously provides a collateral mandate. Within its ten years of existence, the activities of AMCHUD, guided by its multidimensional framework for housing and urban development, has been focused substantively on different aspects of slum upgrading prominent among which is fund mobilization under the aegis of UN-Habitat. The 2015 focus on determining African urban agenda is invariably committed to financial mechanisms. Apparently the focus on African urban agenda is driven by the need to tackle the growing challenge of urbanization on the continent.

At this juncture is it necessary to note that the activities of AMCHUD are under the influence of two notions: financial mechanisms for plan implementation and impressions of the so called rapid urbanization thatchallenge development initiatives. The UN-Habitat cashes in on these notions to assume relevance in the activities of AMCHUD. As such it is noticed that AMCHUD scheme for financing human settlements is invariably extroverted and dependent on external aid notwithstanding the prospects of domestic funding. The International Monetary Fund (IMF) (1993) indicates that over 95 % of investment in any developing country is financed from domestic savings. With regards to impressions of rapid urbanization in Africa, Todaro (1979) observed that institutional database are highly sensitive to error and barely a decade later, Linden (1996) indicated that urbanization in developing countries is not accelerating. Cohen (2003) reaffirmed this position and challenged the assumption of rapid urbanization in developing countries. All along, the impression of rapid urbanization in Africa has been taken with serious reservations and ultimately, Potts (2012) effectively and convincingly debunked this impression with facts and figures. It is not unlikely that the two notions are deliberately deployed as instruments of control.

The NEPAD cities initiative: The mindset and outlook of the NEPAD cities initiative are difficult to reconcile. The UN-Habitat conceived NEPAD city initiative with the mindset of preparing African cities to deliver MDGs hence has their attention focused on slum clearance and installing pro-poor measures in urban development. On the contrary, the outlook of NEPAD cities postulates wide-ranging attributes that aim to achieve world class standards in functional development through the instrumentality of local actors in a participatory planning process. According to AMCHUD (2005: 16), a NEPAD city will be functional, economically productive, socially inclusive, environmentally sound, safe, healthy and secure.

Meanwhile, the NEPAD city initiative is built on the epistemology of imperialism laid by the botched sustainable cities programme, yet its protagonists are still confident that it will strike a 'balanced and symbiotic relationship with its hinterland, one which effectively influences the productivity of agriculture, stimulates strong sub-regional flows of trade, and provides adequate access to domestic and international market' (AMCHUD 2005b: 16). It is not clear how slum clearance heavily criticized for its inadequacy and pro-poor measures, would deliver the vision contained in the conceptual framework of NEPAD cities initiative. This is reflected in the performance of the seven NEPAD cities identified in Phase One which includes: Douala, Bamako, Durban, Lagos, Lusaka, Nairobi and Rabat. Meanwhile UN-Habitat outlined another set of plausible objectives for Phase Two cities.

There are critical impressions that the literature on the NEPAD cities initiative conveys which is highlighted herewith. The fact that the NEPAD cities initiative is a UN-Habitat idea illustrates the redundancy of AMCHUD. Somehow the new partnership arrogates lots of authority to UN-Habitat to initiate and follow through projects while AMCHUD sits back and watches complacently, although where AMCHUD lacks initiative the role of UN-Habitat is justified. So far UN-Habitat procedures and planned action for NEPAD cities seem to fall short of adequate theoretical bases for growth or shaping of the city. Recourse to informal spatial growth, determined through market forces, is evident considering the absence of appropriate spatial development paradigm and the disregard for formal expert knowledge in the midwifery of planning interventions for managing the informal sector of survivalist category as identified in Africa.

The UN-Habitat approach is palliative and scarcely addresses the remote causal factors of informal sectors which impact on the modelling of urban regions and the morphology of cities. There is no gainsaying that the consent to informal spatial growth is more of a submission to the problems of informality, a move that ultimately consolidates the gains of dependent capitalism in Africa. As it were, it works against the rebirth of development for the African renaissance. Overall, a genuine NEPAD city initiative cannot in all good conscience share the mindset and outlook of planning for African cities that submits to the doctrine of informality.

11.2.2 Priority for Action

The priorities for action are drawn from transboundary problems. The problems are herewith recalled as follows

i. Urban spatio-physical expansion;
ii. Suburbanization;
iii. Urban sprawl/slum formation;
iv. Extroverted urban economy;
v. Spatial inequality (i.e. distortions in the urban region);
vi. Informality; and
vii. Introverted urbanization (i.e. city primacy).

11.2.3 Proposed Action Cards

The proposed action cards for integrated regional development are contained in Table 11.2 below. The action cards will be executed at the national level.

11.2.4 Typology of Action

The typology for action is contained in Table 11.3 below. As is expected, planning and standardization attract the highest number of actions followed by management.

Table 11.2 Proposed action cards for integrated regional development in Africa

Priorities for action	Qualitative action cards	Performance indicators
Urban spatio-physical expansion	• Urban renewal scheme, • Vacant land survey, • Urban infrastructure development, • Urban waste management scheme, • Determining urban design standards, • Setting city limits, • Conduct urban growth studies • Research and development (R&D) unit	• Standardized urban design protocols, • Set standards for city limits, • Delineate urban growth boundaries for cities, • Establish space standards, Prepare renewal schemes, land use budget, time budget.
Suburbanization	• Delineate urban growth boundaries (UGBs), • Identification of activity belts, • Housing resettlement schemes, • Industrial park development, • City centre revitalization, • Research and development (R&D) unit	• Establish greenbelts, • Prepare spatial integration plans, • Prepare land use schemes, • Enact interim development order, • Prepare downtown revitalization schemes, • Conduct urban growth studies,
Urban sprawl/slum formation	• Preparation of urban master plans, • Refill and infill development schemes, • Central and peripheral slum renewal schemes, • Renovation of dilapidated urban infrastructure, • Urban quality control • Research and development (R&D) unit	• Vacant land studies, Smart growth activities, • Urban renewal schemes, • Quality urban design, • Affordable housing schemes, Urban management schemes, etc.
Extroverted urban economy	• Farm settlement development, • Development of regional road network, • Preparation of resource management plan, • Market development, • Research and development (R&D) unit • Identification of market regions	• Hinterland farm plantations, • Processing industries, • Regional Development Master plans (regional connectivity schemes), • Growth visions, • Master plans for land resource management, etc.
Spatial inequality	• Hinterland plantation development schemes, • Classification of cities, • Identification of infrastructure grids, • Preparation of Spatial integration plans networking of ICT cities, • Capacity building • Research and development (R&D) unit	• Adopt NUDS, • Establish hierarchical urban system with spatial linkage capabilities, • Regional integration plans, Vision documents that relate modern sector with local economy, etc.

(continued)

11.2 Priority Actions

Table 11.2 (continued)

Priorities for action	Qualitative action cards	Performance indicators
Informality	• Conduct housing demand studies • Mainstream monitoring mechanisms, • Legal reforms for land use management, • Preparation of low-cost housing development scheme • Adoption and domestication of National • Integrated Regional Development Act (NIRDA) • Research and development (R&D) unit	• Reduced growth of informal settlement, squatter settlement, slum locations, etc.
Introverted urbanization	• Master plan: rural growth centres (agrovilles), • Engage in new town development, • Intervention policy for growth pole development, • Preparation of rural housing development scheme • Research and development (R&D) unit	• Reduction of city primacy, • Decentralization schemes,

Source Own construction (2014)

This vindicates the poor planning situation at the moment. Also development and renewal activities drew sizable attention compared with the other typologies of action.

The planning action plays an integrative role. Each planning action comprises of two activities: spatial and investment planning. The spatial planning component will address the distribution of land-use requirements for human systems to facilitate the functional flow of the urban system and ensure spatial regional integration of the action cards in all of the other typologies of action. In doing so, investment planning accompanies the master plans as instrument of flexibility and plan implementation. This permits the relationships and causalities established between informality and planning to come into play for project development and for confirming land-use functions vis-à-vis inclusiveness however all within the spatial framework of the master plan. The existing planning initiatives, which are mainly project/sector development-oriented, will forebear the investment planning.

Table 11.3 Typology of action

S/no	Typology of action	Action cards
1	knowledge	• set up research and development (r&d) units • conduct housing demand studies • conduct urban growth studies • vacant land survey,
2	planning/standardization	• master plan for rural growth centres (agrovilles), • preparation of spatial integration plans (sips), • hinterland plantation development schemes, • preparation of resource management plan, • preparation of urban master plans, • setting city limits, • determining urban design standards, • urban waste management scheme,
3	development	• engage in new town development, • market development, • development of regional road network, • farm settlement development, • industrial park development, • urban infrastructure development,
4	institutional	• legal reforms for land use management,
5	environmental	• urban quality control
6	monitoring	• mainstream monitoring mechanisms,
7	renewal	• renovation of dilapidated urban infrastructure, • central and peripheral slum renewal schemes, • refill and infill development schemes, • city centre revitalization, • urban renewal scheme,
8	housing	• preparation of rural housing development scheme • preparation of low-cost housing development scheme • housing resettlement schemes,
9	management	• adoption and domestication of national integrated regional development act (nirda), • identification of market regions • networking of ict cities • identification of infrastructure grids, • classification of cities, • identification of activity belts, • delineate urban growth boundaries (ugbs), • capacity building
10	policy	• intervention policy for growth pole development,

Source Own construction (2014)

11.3 Implementation Strategies

The implementation phase of neo-mercantile planning is vulnerable to funding mechanisms that are driven by international funding institutions. This signals dependence which in important ways weakens the new paradigm but much depends on the resolve of Africa and the Diaspora to chart her destiny. But unfortunately the attitude of resignation to trends is overwhelming. This does not augur well for constructive change. The strong point of neo-mercantile planning is that the rationale for its application is getting more evident and the literature is gradually responding to the need for the reversal of trends especially in Africa. Neo-mercantile theory is an integrated approach of reversing trends in planning. Planning elements are interlinked thus requiring holistic review to secure positive result. It is not unlikely that thematic reviews would appear but presumably that would be the bane of reversing trends in planning in Africa.

11.3.1 Institutional Requirements and Implementation Processes

The implementation process explains the sequence of central coordination of decentralized planning activities driven by the action cards. The following institutions are identified to play major roles in the implementation process at continental level:

  i. African Union
 ii. African Ministerial Conference on housing and urban development (AMCHUD)
iii. NEPAD
 iv. National NEPAD subsidiaries
  v. Commonwealth Association of Planners (CAP)
 vi. African Association of Planners (AAP)
vii. African Association of Planning Schools (AAPS)
viii. National Planning Commissions
 ix. Planning education facilities (Universities, research institutes, tertiary institutions, etc.)
  x. Regional Networks including:

- Southern African Development Community (SADC),
- Economic Community of Central Africa States (ECCAS),
- West African Economic and Monetary Union (WAEMU),
- Community of Sahel-Saharan States (CEN-SAD),
- Economic Community of West African States (ECOWAS),
- Common Market for Eastern and Southern Africa (COMESA),
- Intergovernmental Authority on Development (IGAD)
- Economic Commission of Africa (ECA),
- Convention to Combat Desertification (CCD)
- Comprehensive Africa Agricultural Development Programme (CAADP),
- African Peer Review Mechanism (APRM)

Table 11.4 Action cards implementation process

Stages (zones)	Cognate action cards	Major agency(s)	Other agencies
Training	• Capacity building	• National Planning Commissions,	• African Association of Planning Schools (AAPS), • Planning education facilities (Universities, research institutes, tertiary institutions, etc.)
Mobilization	• Intervention policy for growth pole development, • Adoption and domestication of National Integrated Regional Development Act (NIRDA) • Legal reforms for land use management,	• African Ministerial Conference on housing and urban development (AMCHUD)	• Regional Networks • Regional Development Authorities
Logistics	• Delineate urban growth boundaries (UGBs), • Identification of activity belts, • Classification of cities, Identification of infrastructure grids, • Identification of market regions, • Networking of ICT cities	• National NEPAD subsidiaries	• Regional Networks • UN-Habitat Regional Development Authorities
Plan generation	• Master plan for rural growth centres (agrovilles), • Preparation of Spatial integration plans (SIPs), • Hinterland plantation development schemes, • Preparation of resource management plan, • Preparation of urban master plans, • Setting city limits, • Determining urban design standards, • Urban waste management scheme, • Preparation of rural housing development scheme • Preparation of low-cost housing development scheme	• National Planning Commissions, • New Town Development Authorities	• Commonwealth Association of Planners (CAP), • African Association of Planners (AAP). • Regional Development Authorities

(continued)

11.3 Implementation Strategies

Table 11.4 (continued)

Stages (zones)	Cognate action cards	Major agency(s)	Other agencies
Sector programmes	• Housing resettlement schemes • Renovation of dilapidated urban infrastructure, • Central and peripheral slum renewal schemes, • Refill and infill development schemes, • City centre revitalization, • Urban renewal scheme,	• NEPAD • New Town Development Authorities	• Commonwealth Association of Planners (CAP), • African Association of Planners (AAP). UN-Habitat
Intervention	• Urban quality control • Engage in new town development, • Market development, • Development of regional road network, • Farm settlement development, • Industrial park development, Urban infrastructure development,	• National Planning Commissions, • New Town Development Authorities	• Regional Networks • Regional Development Authorities
Monitoring	• Mainstream monitoring mechanisms,	• African Union Regional Development Authorities	• Regional Networks • UN-Habitat Commonwealth Association of Planners (CAP), • African Association of Planners (AAP).
Research	• Setup Research and development (R&D) units • Conduct housing demand studies • Conduct urban growth studies • Vacant land survey	• African Ministerial Conference on housing and urban development (AMCHUD), • New Town Development Authorities • UN-Habitat	• African Association of Planning Schools (AAPS) • Planning education facilities (Universities, research institutes, tertiary institutions, etc.)

Source Own construction (2014)

   xi. New Town Development Authorities
  xii. Regional Development Authorities
- Niger Basin Development Authority
- Chad Basin Development Authority

xiii. UN-Habitat

These institutions have definite roles to play in the 38 action cards that are grouped into eight stages (zones) of plan implementation processes as shown in Table 11.4 below. The agencies will cooperate with each other and play their statutory roles as they relate to the functions of the action cards. However, AMCHUD will always provide the necessary coordination of activities of the various agencies and the necessary liaison with AU and the different national governments.

11.3.2 Manpower Requirements

The status of technical and professional manpower in the various agencies is not clear at the moment. However, increased manpower will be required primarily for the following technical and professional experts: planners with specialization in urban design and transportation, environmental economists, sociologists, environmentalists, landscape architects, lawyers and land surveyors. These as well as other allied professionals concerned with regional development represent manpower requirements, but details and specificity on actual requirement will entail a study on current manpower inventories especially those of core urban planners.

11.3.3 Financial Mechanisms

The African Union shall oversee the funding of central coordination. Financial arrangements at country level will involve four major stakeholders, namely the Government of the African country, the Corporate Private Sector (including multinational organizations), Foreign Partners and Donor Agencies. The full financial model will require the deliberation of these stakeholders in a consultative forum. However the ground rules of spatial development set out in the spatial integration network shall not be subject to aid conditionality.

11.3.4 Legal Reforms

Legal reforms are incidental for the installation of effective planning in Africa and more so with the prospects of significant paradigm shifts in the process. This was the case with EU countries when they had to engage the Planning and Compulsory Purchase Act 2004 preparatory to the launch of European Spatial Development Perspective (c.f.6.3). The legal reform anticipated for Africa shall support visionary planning practice which is the strong point of the neo-mercantile planning paradigm. The new planning paradigm instrumentality offers a centralized system of managing decentralized planning (c.f. 10.1.2). It allows planning rationality and market forces to work complementarily to pursue economic growth in the context of shaping the city and regional spatial systems.

11.3 Implementation Strategies

Therefore the reforms will rework the tools of planning to strengthen the developmental role of the planning system. It will promote professionalism but provide strategic frameworks for effective engagement with stakeholders. It will encourage plan-led development in the context of private investment. In other words, financial mechanisms will not usurp integration. Overall the reforms will deal with issues outlined earlier (c.f. 10.1.5.1).

11.3.5 Monitoring Measures

Statutory monitoring by stakeholders is envisaged and should be mainstreamed in the planning and implementation process. The African Peer Review Mechanism (APRM) of the African Union is expected to play a major role. Observatories should be erected to track development and chart the progress in plan implementation which can be presented in the form of GIS database. In other words, GIS databases of spatial systems shall serve as monitoring instrument and thus be prepared for the spatial integration network in each country.

11.4 Calendar of the Action Plans for Spatial Regional Integration in Africa

The plan implementation process as shown in Table 11.4 above is calibrated into eight stages. These stages are allocated to three phases as shown in Table 11.5 below. The implementation of the short, medium and long-term categories will be operated concurrently within a time frame of fifteen years (2016–2030) (using 2016 as base year) and the mission ahead is contained in the provisions of quantitative performance indicators.

The training and mobilization activities in the first phase (short term) contain four action cards, mainly for central coordination and geared towards creating enabling environment for take-off. It is proposed that African Ministerial Conference on housing and urban development (AMCHUD) will house the central coordinating body. This body will play an active role in securing political will for the transition and this precedes the activation of the action cards. The action cards require light investment but demand lots of commitment and determination. This is why the capacity-building segment will be located in the process of psyching the mindset of stakeholders properly through motivational talks and the legal reforms will empower and share responsibilities to stakeholders. The curriculum for capacity building will draw from issues recommended for harmonizing the theoretical framework for spatial planning (c.f. 10.1.5).

The second phase (medium term) contains 22 action cards. This phase domesticates the enabling measures at the national level. It focuses on elaborating the

Table 11.5 Phases of the action plan

S/no	Phases	Time frame	Implementation process	Quantitative performance indicators
1	Short term (in 5 years)	2016-2020	• Training • Mobilization	• Reduction of semi-professional man power ratio by 50 %, • Completion of legal reforms, • Full reform of policy framework for spatial development,
2	Medium Term (in 10 Years)	2016-2025	• Logistics • Plan generation • Sector programmes	• Application of urban growth boundaries (UGBs) in at least 75 % of national and provincial cities, Complete classification of National and Provincial cities, • Full activation of activity belts, infrastructure grids at National, Provincial and Local levels, Completion of national market region delineation, Completion of plan preparation for spatial development at Provincial and local level, Completion of renewal schemes for 75 % of slum locations.
3	Long Term (in 15 years)	2006-2030	• Intervention • Monitoring • Research	• Setting-up of functional regional development authorities at Provincial level, • Setting-up of functional urban observatories in all market regions, • Setting-up of functional research institute for spatial development.

Source Own construction (2014)

planning instruments, both provisional and regulatory. It requires heavy investment for professional services envisaged for the new methodology of preparing plans to remodel the status quo of spatial systems. To this end the planning system will require capacity-building operations in terms of manpower upgrading to overcome mediocrity. Again the training curriculum will draw on the recommendations in Table 10.1 (c.f. 10.1). The public sector will bear the bulk of investment at this stage.

The third and last phase (long term) has twelve action cards mainly geared towards direct intervention linked with monitoring and research activities. Heavy investment is anticipated here but from the private sector for development activities. Monitoring activities are expected to move in tandem with the rate of growth and conducted with the monitoring instrument proposed. Monitoring activities will provide feedback for research and development. The development of planning curricula for university education will be modified accordingly to reflect the ethos of neo-mercantile planning theory.

11.5 Conclusion

The chapter sets out with a visioning process for domesticating the newly identified neo-mercantile planning theory in Africa. The visioning process starts with the identification of priority transboundary problems in Africa on the bases of which development objectives—short, medium and long-term objectives—were determined to verify the vision statement. A substantive vision statement was provided to act as a guide for the proposed priority actions. An assessment of current action was given on the basis of which priorities for action were determined, then followed by the projection of action cards that were subsequently processed into typologies for action to facilitate implementation strategies. The implementation strategy made provision for institutional requirement and implementation processes involving manpower requirements, financial mechanisms, legal reforms and monitoring measures. Provision was also made for a calendar of the action plan for spatial regional integration in Africa.

Overall, the peculiar circumstances surrounding Africa require a vigorous pursuit of economic growth in tandem with spatial integration in planning interventions. At least it is reasonably clear that following current trends in planning cannot achieve the foregoing objective. At the same time, under the overarching influence of a neoliberal ideological perspective, summoning enough political will to effect attitudinal change is a herculean task. However with the right mindset and outlook, which is likely to come from coordinated awareness campaigns, receptive attitude towards neo-mercantile planning could be mustered. Planners, both consultants and administrators, are expected to rise to the occasion and reinforce their theoretical arsenal through targeted research in planning theory and planning protocol. It is anticipated that African planning schools and research institutes will provide research support. Immediate action is required because the need to halt the influence of neoliberal values in the development of curriculum for planning education in African planning schools cannot be over emphasized.

The change of attitude of the political class is considered critical for take-off. Granted that this input contributes the theoretical foundations for change, developing realistic frameworks for concrete action are strategic short-term objectives that are surmountable. Minimal institutional changes are required however a little bit of legal reform is required. Already requisite political and economic reforms are in place and represented in NEPAD initiative. The **AU** naturally leads the institutional base and provides the platform to negotiate the political will that shall lead up to the anticipated legal reform. However, there is a need to re-engineer the working institutions of **AU** especially capacitating AMCHUD and reposition it for its new role in delivering territorial planning in Africa. Also setting-up a research institute as support structure under **AU** is in the agenda to domesticate the proposed neo-mercantile paradigm.

The neo-mercantile planning paradigm is programmed to deliver integrated development. The type intended through NEPAD initiative for Africa, which is not measured in terms of the extent MDG(s) objectives are achieved, but the extent to which the stated objectives of NEPAD ideology are achieved. The objectives under reference aim at the delivery of Africa from dependent capitalism. In practical terms, this means introverting the already extroverted economy of urban Africa through planning interventions. This will impact on changes in African space economy for the delivery of the African renaissance.

The African model of domesticating neo-mercantile planning theory may not apply directly in other developing economies abroad. This is understood because local scenarios differ. But however different the local conditions are, the intrinsic similarity is that these economies share the prospects of securing integration in the development of their spatial systems. The instrumentality of neo-mercantile planning theory is postulated because it is realistic, malleable and committed to reverse trends that sustain an imperial space economy in the so-called source regions worldwide.

"A theory must be tempered with reality"

–Jawaharlal Nehru

Indian politician (1889–1964)

References

Amuchud (African Ministerial Conference on Housing and Urban Development) (2005) African Cities Driving the NEPAD Initiative: an Introduction to the NEPAD Cities Program. Durban, South Africa. 31 Jan 2012
Bond P (2002) Unsustainable South Africa: environment, development and social protest. Merlin Press, University of Natal Press, London, Pietermaritzburg
Bond P, Dor G (2003) Neoliberalism and poverty reduction strategies in Africa. Discussion paper for the Regional Network for Equity in Health in Southern Africa (EQUINET March 2003)
Cohen B (2003) Urban growth in developing countries: a review of current trends and a caution regarding existing forecasts. World Dev 32(1):23–51
International Monetary Fund (IMF) (1993) World economic outlook. IMF, Washington, DC
Linden E (1996) The exploding cities of the developing regions. Foreign Aff 5(1):52–65
Nabudere DW (2003) Towards a new model of production—an alternative to NEPAD. 14th Biennial Congress of AAPS. Durban, South Africa. www.mpai.ac.ug. Date of access 8 July 2011
Potts D (2012) Whatever happened to Africa's rapid urbanization. Africa Research Institute. Available at: www.researchgate.net/.../254456370. Date of access 9 Nov 2015
Richard J (2002) Smart partnerships for African development: a new strategic framework. (PDF Special Report No 88. Available at http://www.usip.org/pubs/specialreports/sr88.pdf page 2 of 11). Date of access: 9 October 2007
Tawfik RM (2008) NEPAD and African development: towards a new partnership between development actors in Africa. Afr J Int Aff 11(1):55–70p
Todaro MP (1979) Urbanization in Developing Nations: trends, prospects, and policies. Center for Policy Studies, Working Paper Series. The Population Council. (Reprinted in Ghosh, 1984)

Appendix A

A.1 Matrix of Descriptive Statistics of Related Themes Considered for Measuring Formal (Form-Based) Planning

S/No	Questions (themes)	Perceptions			
		+1	0	−1	M
		%	%	%	%
1.	In an IDP planning approach is the urban environment regarded as the physical shape of the city?	39.4	30.3	27.2	3.0
2.	Does the IDP function more with land use schemes than PGDS, for instance?	21.2	27.3	42.4	9.1
3.	Do you think IDP plan preparation is essentially a technical activity?	45.5	27.3	27.3	–
4.	Does the IDP serve as an instrument for the development of empty spaces and the renovation of degraded neighbourhoods within the urban environment?	30.3	30.3	36.4	3.0
5.	Was adequate consideration given to the use of professional planning expertise for the preparation of the IDP?	18.2	33.3	42.4	6.1
6.	Does IDP plan preparation consider regional plan classification (such as the urban region; functional region; planning region; physical formal region; economic formal region, etc.)?	48.5	36.4	15.2	–
7.	Is the IDP an instrument for the implementation of the national development corridor strategy?	48.5	33.3	18.2	–
8.	Does IDP preparation consider the connectivity responsible for the functional flow of activities within the urban region?	51.5	36.4	12.1	–
9.	Do you really think that the IDP can effectively repackage the economic fundamentals of municipalities in South Africa?	48.5	39.4	12.1	–
10.	Does the IDP follow any policy guideline for urban development?	57.6	21.2	12.1	9.1
11.	Do IDP activities recognize and apply the spatial aspect of urban growth (such as qualitative, quantitative, structural growth, etc.)?	36.4	39.4	18,2	6.1
12.	Does the planning system consider environmental factors along with culture, value systems, activity systems and their distribution in space as attributes of the urban environment?	42.4	21.2	12.1	24.2

(continued)

(continued)

S/No	Questions (themes)	Perceptions			
		+1	0	−1	M
		%	%	%	%
13.	Does the use of planning standards prevail in South African planning system?	*39.4*	36.4	3.0	21.2
14.	Are greenbelts or other management techniques such as urban service limits, urban growth boundaries (UGB), urban development boundaries (UDB) used to manage urban growth in South Africa?	*51.5*	24.2	–	24.2
15.	Do long-term objectives drawn from a defined mindset and outlook determine city planning and development in South Africa?	*39.4*	36.4	6.1	18.2
16.	Are sanitation and urban quality integrated in South African planning system?	*45.5*	18.2	15.2	21.2
17.	Does planning in South Africa adopt definite spatial measures or standards to shape the city?	*36.4*	30.3	15.2	18.2
18.	Is South Africa inclined to design-oriented approach to planned development?	18.2	*39.4*	24.2	18.2

Note The highest score italicized in each row represents the preferred perception for each question
Note +1 = Yes; 0 = Moderate; −1 = No
Source Own construction in collaboration with Statistical Consultation Service, NWU 2013

A.2 Split-Cell Frequency Distribution of Preferred Perception of Planning Practice

Perception	Frequency	%	Category	Frequency	%	Ratio
Yes	22	61.1	Formal	13	36.1	1.4
			Pragmatic	9	25.0	
Moderate	6	16.7	Formal	2	5.6	0.5
			Pragmatic	4	11.1	
No	8	22.2	Formal	3	8.3	0.6
			Pragmatic	5	13.9	
Total	36	100		36	100	

Source Own construction in collaboration with Statistical Consultation Service, NWU 2013

A.3 Split-Cell Frequency Distribution of Preferred Perceptions Based on Column Percentages of Descriptive Statistics on Planning Practice

Matrix	Administrators					Politicians					Academics					Consultants				
	FREQ	%	C	FREQ	%	FREQ	%	C	FREQ	%	FREQ	%	C	FREQ	%	FREQ	%	C	FREQ	%
Yes	17	47.2	F	8	22.2	16	44.4	F	7	19.4	9	25.0	F	6	16.7	12	33.3	F	4	11.1
			P	9	25.0			P	9	25.0			P	3	8.3			P	8	22.2
Moderate	11	30.5	F	6	16.7	2	5.5	F	2	5.6	15	41.7	F	6	16.7	18	50.0	F	10	27.8
			P	5	13.5			P	-	-			P	9	25.0			P	8	22.2
No	8	22.2	F	4	11.1	18	50.0	F	9	25.0	12	33.3	F	6	16.7	6	16.7	F	4	11.1
			P	4	11.1			P	9	25.0			P	6	16.7			P	2	5.6
Total	36	100		36	100	36	100		36	100	36	100		36	100	36	100		36	100

Note F = Formal (form-based) planning; P = Pragmatic (non-form-based) planning; C = Category
Source Own construction in collaboration with Statistical Consultation Service, NWU 2013

A.4 Split-Cell Frequency Distribution of Preferred Perceptions Based on Row Percentages of Descriptive Statistics on Planning Practice

Matrix	Administrators					Politicians					Academics					Consultants				
	FREQ	%	C	FREQ	%	FREQ	%	C	FREQ	%	FREQ	%	C	FREQ	%	FREQ	%	C	FREQ	%
Yes	21	58.3	F	12	33.3	23	63.8	F	13	36.1	11	30.6	F	7	19.4	16	44.4	F	7	19.4
			P	9	25.0			P	10	27.8			P	4	11.1			P	9	25.0
Moderate	9	25.0	F	4	11.1	3	8.3	F	1	2.8	18	50.0	F	8	22.2	17	47.2	F	10	27.8
			P	5	13.9			P	2	5.6			P	10	27.8			P	7	19.4
No	6	16.7	F	2	5.6	10	27.8	F	4	11.1	7	19.4	F	3	8.3	3	8.3	F	1	2.8
			P	4	11.1			P	6	16.7			P	4	11.1			P	2	5.6
Total	36	100		36	100	36	100		36	100	36	100		36	100	36	100		36	100

NOTE F = Formal (form-based) planning; P = Pragmatic (non-form-based) planning; C = Category
Source Own construction in collaboration with Statistical Consultation Service, NWU 2013

A.5 Split-Cell Frequency Distribution of Preferred Perceptions Based on Mean Values of Descriptive Statistics on Planning Practice Per Category of Respondents

Matrix	Administrators					Politicians					Academics					Consultants				
	FREQ	%	C	FREQ	%	FREQ	%	C	FREQ	%	FREQ	%	C	FREQ	%	FREQ	%	C	FREQ	%
Yes	12	33.3	F	9	25.0	14	38.9	F	8	22.2	5	13.9	F	3	8.3	9	25.0	F	4	11.1
			P	3	8.3			P	6	16.7			P	2	5.6			P	5	13.9
Moderate	15	41.7	F	6	16.7	11	30.6	F	6	16.7	13	36.1	F	9	25.0	18	50.0	F	11	30.6
			P	9	25.0			P	5	13.9			P	4	11.1			P	7	19.4
No	9	25.0	F	4	11.1	11	30.6	F	4	11.1	18	50.0	F	6	16.7	9	25.0	F	3	8.3
			P	5	13.9			P	7	19.4			P	12	33.3			P	6	16.7
Total	36	100		36	100	36	100		36	100	36	100		36	100	36	100		36	100

NOTE F = Formal (form-based) planning; P = Pragmatic (non-form-based) planning; C = Category
Source Own construction in collaboration with Statistical Consultation Service, NWU 2013

A.6 Matrix of Ratios of Positive Perceptions Per Category of Respondent

Matrix	Administrators				Politicians				Academics				Consultants			
	VSR ≥1.0	SR ≥0.5	WR ≥0.0	N ≤0.0	VSR ≥1.0	SR ≥0.5	WR ≥0.0	N ≤0.0	VSR ≥1.0	SR ≥0.5	WR ≥0.0	N ≤0.0	VSR ≥1.0	SR ≥0.5	WR ≥0.0	N ≤0.0
Column percentages		0.9				0.8			2.0					0.5		
Row percentages	1.3				1.3				1.7					0.8		
Mean values	3.0				1.3				1.5					0.8		

NOTE VSR = Very strong resilience; SR = Strong resilience; WR = Weak resilience; N = Negative
Source Own construction in collaboration with Statistical Consultation Service, NWU 2013

A.7 Summary of Research Findings

S/No	MCA analysis	Reference	Findings
1.	Compliance with the new theoretical framework in Africa	c.f. 6.5.1	• Multidimensional planning perspective exists in Africa • Informal planning instrument is used for managing spatial development in Africa • Planning outlook is diffused in Africa • Spatial planning is not form-based in Africa • There exists nearly a mean relationship between theoretical and analytical framework in spatial planning in Africa
2.	Compliance with the new theoretical framework in planning initiatives in selected African countries	c.f. 6.5.2	• The synthesis of data generated as contained in Table 6.8 indicates that IDP initiative in South Africa is more compliant to neoliberal planning theory as indicated in Fig. 6.2 above. Figure 6.3 confirmed that the initiatives studied cumulatively indicate the dominance of strong compliance although Fig. 6.4 indicates that the compliance surface undulates • Market forces are the dominant determinant factor for land use management in Africa • The participatory process is invariably consultative in Africa • Project planning defines planning framework in Africa
3.	Own assessment of planning initiatives in selected African countries using **4As** criteria	c.f. 6.6.1	• The IDP initiative exhibits the best disposition towards compliance with the new spatial planning theoretical framework that has developed since the 1980s • Planning initiatives in Africa are generally weak spatial planning instruments. They are all weak in spatial integration and somehow strong in resource mobilization • Planning initiatives in selected African countries cumulatively indicate very low percentages of strong capacity to deliver spatial regional integration in the continent as indicated in Fig. 6.6 • The compliance of planning initiative with neoliberal planning is on a higher platform in South Africa and Egypt compared with the other initiatives that were studied

(continued)

(continued)

S/No	MCA analysis	Reference	Findings
4.	Compliance with the new spatial planning theoretical framework in selected African countries	c.f. 6.7	• The level of mean performance at roughly 2.5 % is above the average threshold of 2.4% anticipated per variable for a total of 42 variables (options) considered. Thus the distribution of performance levels is healthy • The bulk of issues beneath the mean average line are form-based issues and those above are mainly developmental economics, informal expertise and participatory issues • The most compliant country to the new theoretical framework is South Africa and the least is DRC
5.	Desktop case studies of select IDPs in South Africa	c.f. 6.8.3	• The performances of IDP frameworks vary but they all maintain a positive level of compliance with the new theoretical framework • The relationship between existing and desired practice is positive although it varies in relation with compliance with principles • Strong positive relationship exists across board between principles and desired practice • Weak relations are at a low ebb and more pronounced in the relationship between practice and desired practice • The capacity of IDPs to achieve spatial regional integration is generally below average although comparative advantages exist between IDPs
6.	SWOT analysis of IDP/SDF in Tlokwe, Matlosana and Rustenburg municipalities	c.f. 6.8.4	• The IDP is a participatory (neoliberal) planning instrument • The IDP as instrument for spatial planning is contestable • Not all IDP documents are visionary • The IDP conceptually lacks an integrative element. There is need to mainstream integration • The SDF is a misplaced strategy because the SDF is conceptually invalid under an IDP culture • The IDP lacks authority in planning practice

(continued)

Appendix A

(continued)

S/No	MCA analysis	Reference	Findings
7.	Own assessment of IDPs using **4As** criteria	c.f. 6.8.5	• The pattern and quality of IDP documents vary • IDP planning is potentially weak and vulnerable as a spatial planning instrument • IDP documents are fairly well-related to the IDP theoretical framework • The IDP is more advanced in theory than in practice • Patterns of relationships between principles, practice and desired practice are not identical but fairly similar within and across the municipalities • Levels of relationships between principles, practice and desired practice within and across municipalities are consistently positive
8.	Empirical data on IDP/SDF local municipalities in South Africa (Empirical case studies)	c.f. 6.9	• IDP practice is generally perceived to reinforce the tenets of new spatial planning principles • Administrators and politicians discretionally are more inclined than academics and consultants to perceive positive compliance of IDP practice with spatial planning principles • Mutual perceptions indicate that only administrators feel strongly about the positive compliance of IDP practice with spatial planning principles while politicians think otherwise and academics and consultants maintain a moderate position • Mean average perceptions leave the politicians to maintain a precarious support for the compliance of IDP practice to spatial planning principles while academics gets more pessimistic • There is no significant relationship between the categories of perceptions as measured by the 'Effect size' calculated for planning criteria investigated and • Overall the perception of politicians is more sensitive considering its fluctuation while that of academics and perhaps consultants is more stable and consistently not in favour of the compliance of IDP practice to

(continued)

(continued)

S/No	MCA analysis	Reference	Findings
			spatial planning principles. The perception of administrators shares some measure of stability, however, in favour of positive compliance of IDP practice to spatial planning principles
9.	IDP/SDF interview summary	c.f. 6.10	• Irrespective of compliance with the new theoretical framework there exists a consensus among all categories of respondents that the IDP initiative is potentially a weak planning instrument. In other words it is not achieving its theoretical role especially integration • The views of academics, consultants and politicians are symmetrical in collaborating the weakness identified • The views of administrators differ from the symmetry shared by the other categories of respondents and tended to be relatively optimistic due to their positive appraisal of IDP as a spatial planning methodology with potentials to shape the city and subsequently the space economy • The four categories of respondents are consistent in their views that the IDP at the moment exhibits a negative in securing integration and indicates a positive as to remote causal factors or 'silos' of development given its disposition towards sectoral planning and • Notably the politicians are more incisive (and somehow pessimistic) about the IDP, thus signalling their role as potential content drivers in planning

Source Own construction (2013)

Glossary

African city Not all cities in Africa are indigenous cities. In fact, it can be argued that indigenous African cities are technically extinct. Therefore, the phrase 'African cities' is used loosely in the literature to refer to hybrid cities that do not necessarily derive from indigenous values, attitudes and institutions; hence they are not responsive to indigenous enterprise and culture. Africa lost its heritage of city development and this is evident in the epistemology of its civilization. But civilization is indeed the culture of cities, therefore, African cities are perceived in this discourse as representing African civilization

African renaissance The concept of 'African renaissance' under the influence of South Africa was popularized with the inception of the NEPAD initiative. In its original form it indicated regional integration in political and economic terms. Since its conception general discussions have tended to associate the African renaissance with the economic rebirth of Africa. This discourse identifies with this imperative but goes further to link the African renaissance with the reworking of African space economy. This implies the inclusion of the space dimension as a critical element in managing resource economy in contemporary Africa

Agro-politan development This development approach pioneered by Friedman and Douglass in 1978 has its roots in the paradigm of territorial development. It is progressively conceived as a spatial framework for rural development oriented to human needs with a more equitable distribution of economic benefits and direct movement of local people in the process of development and growth. This is based on the activation of rural people, agriculture and resources. This discourse accepts this definition without reservation; however, an assets-based analysis for its application is emphasized

Agro-villes The concern for food security especially drew attention to traditional hamlets that exist as food baskets but that hitherto has been terribly neglected. Hamlets have proved to be resilient under changing economic, social and political conditions. However, in their present disposition they cannot continue to service increasing food requirements. The transition of the hamlets to enable this function elicited the agro-ville concept, which seeks to increase food production through the provision of a functional base for all categories of potential agriculturists, particularly food-crop producers. It is in this context that the agro-ville concept is applied in this discourse. At present the term agro-ville is used in Pakistan to refer to small and medium towns from the Growth Centre perspective, particularly in the 1970s and mid-1980s

Development Friedmann (1972: 84) indicates that development can 'occur if growth is allowed to pass through a series of successive structural transformations of the system'. He further submits that development is 'an innovative process leading to the structural transformation of social systems'. In Africa his second submission applies, development is an innovative process than a growth process. In recent times innovations parachute into the spatial system mostly through neoclassical investment mechanisms without impacting the social system. This is not acceptable.
Preferably development is a growth process that relates to the unfolding of the creative possibilities inherent in society. It is therefore perceived to connect with Edwardo's (1990) idea of civilization which implies that the culture of cities is built on indigenous values, attitudes and institutions

Growth Friedmann (1972: 86) posits that growth refers to an expansion of the system in one or more dimensions without a change in its structure. From the spatio-physical perspective this correlates strongly with the qualitative urban growth concept in which a new unit is introduced into an urban system. Also quantitative, structural and smart growth manifest variously. The operational growth in this discourse is structural growth which refers to growth of any complex structure that is associated with changes in form. In this case 'the growth process involves changes in the relationship of the parts' (Edwardo 1990: 109)

Planning concept Current trends in which planning is regarded as an event and not as an activity are not acceptable in this discourse. Acceptable definitions of planning abound but those posted in the net by Ravi Business Studies are apt and are used in this discourse. This internet material indicates that planning means deciding in advance what is to be done, when, where, how and by whom it is to be done. Planning bridges the gap from where we are to where we want to go. It includes the selection of objectives, policies, procedures and programmes from among alternatives. A plan is a predetermined course of action to achieve a specified goal. It is an intellectual process characterized by thinking before doing. It is an attempt on the part of manager to anticipate the future in order to achieve better performance. Planning is the primary function of management

Spatial planning Trends of spatial planning losing its essence as a tool for determining the use of space are gaining momentum. This is as a result of attempts to enlarge its content to address cross-cutting, rather than concentrating on core planning issues. Hitherto at its inception spatial planning dealt with the management of land use change (Todes et al. 2010: 416).This discourse maintains fate with its concern for the core issues of space-activity relationships in managing land use change. However, this concern is linked with the distribution of resource utilization in space as represented in territorial planning sometimes referred to as territorial cohesion (see Faludi 2005)

Spatial system According to Friedmann (1972) a spatial system is a territorially organized social system. He further explains that spatial systems are integrated through a given structure of authority-dependency relations maintained partly by a belief in the legitimacy of the relation itself and partly by coercion. Therefore Friedmann's perception of spatial systems is based on authority structures. Edwardo (1990), on the other hand, explained that human systems in planning cannot be adequately explained if they are not related to space. This discourse is inclined towards Edwardo's relation of human systems to space. Hence a spatial system is perceived as a territorial concept that is expressed in geographic space with human elements engaging in a functional flow of activities

Urban form Urban form is a growth-dependent variable. It is the function of the factors of urban land use distribution, urban growth patterns, and urban activity systems. Urban form will evolve in a suitable manner as size increases and size and form limit and determine one another, etc. The consideration of critical mass lead to the alternative and perhaps more appropriate nomenclature of urban form as community form. The community form as a sustainable physical spatial-form is a growth-dependent variable with factors of change characterized in land use as mixed development, in social imperatives and community as critical mass or heterogeneous nucleation, in economy as urban employment and in transit as walk-ability, among others. These accredited perceptions derive from submissions of Edwardo (1990)

World-system The world system is a variant of the neo-Marxist approach of viewing the global economy. The world system in a neo-Marxist perspective, which explains the mechanism through which growth and strength of the core regions of global economy are made possible by the exploitation of the rest of the world (see Portes and Walton 1981; Castells and Portes 1989). The world system is therefore built on the core-periphery principles which the institutionalist school upholds for the delivery of globalization. In practical terms at the local level this finds expression in the dichotomy of urban and rural economies and most times there is a backward linkage driven by informality. The proliferation of informality is underway in a neoliberal dispensation. This is at variance with the mindset of this discourse. Hence, the world system is perceived as the bane of regional integration in developing economies and in Africa in particular

Neo-mercantilism Neo-mercantilism is founded on the use of control of capital movement and discouraging of domestic consumption as a means of increasing foreign reserves and promoting capital development. This involves protectionism on a host of levels: both protection of domestic producers, discouraging of consumer imports, structural barriers to prevent entry of foreign companies into domestic markets, manipulation of the currency value against foreign currencies and limitations on foreign ownership of domestic corporations

Neoliberalism Neoliberalism is a philosophy in which the existence and operation of a market are valued in themselves, separately from any previous relationship with the production of goods and services, and without any attempt to justify them in terms of their effect on the production of goods and services, and where the operation of a market or market-like structure is seen as an ethic in itself, capable of acting as a guide for all human action, and substituting for all previously existing ethical beliefs

Index

A
AbdouMaliq Simone, 187
Accumulation
　regimes of, 38
Activity(ies)
　agricultural, 51
　monitoring, 73
　visionary, 117
Actors, social, 40
Administration, 5
Administrators
　colonial, 37
　planning, 37
African civilization, traditional, 184
African countries, Francophone, 130
African renaissance, 170
　as the object of planning, 163
　principles of, 177
African Union, 20
Agencies, administrative, 64
Agents, external, 114
Analysis, SWOT, 138, 158
Anglophone, 130
Angola, 30, 54
Approach
　conventional, 115
　design-oriented, 111
　interventionist, 110
　world systems, 97
Area, hinterland
　fringe, 118
Areas
　catchment, 105
　fringe, 59
　informal, 58
　peri-urban, 59
　peripheral, 59
　residential, 98
Argument

　growth, 101
　major, 27
ASEAN, 129
Assessment, development, 89
Assumption
　educated, 27
Attitude, 68
　repressive, 55
AU, 56
Authority, decision-making, 128

B
Baeten, 91
Basil Davison, 186

C
Capacity
　endogenous, 97
　integrative theoretical, 115
Capitalism
　dependent, 45
　economy, 12
　fettered, 81
　international, 97
　peripheral, 45
Centers
　growth, 12
Century, 15
Change, urban
　physical, 101
Chukuezi, 3
Cities, Latin America, 126
Cities, worldwide
　vatican, 187
City
　primacy, 53
　resilient, 4
Civilization
　African, 17

Civilization (*cont.*)
    traditional, 84, 139, 171
    tropical, 169
Classification, regional, 103
Classless, 82
Collaboration, inter-organizational, 131
Communities, gated, 90
Community
    homogeneous and specialized, 47
Concept
    efficency wage, 106
    product cycle, 107
    regional rotation, 107
    retchet effect, 107
    spatial-biases, 107
    take-off period, 107
    theoretical, 8
    trickling-down effect, 106
    verdoom effect, 106
Consultancy, planning services, 56
Consultants, 62
    foreign, 56
    national, 62
Context
    of imperialism, 27
Control
    development, 73
    planning, 73, 115
Corridors
    development, 55
        national, 67
    logistic, 55
    national, 55
    transport, 55, 98
Cosmology, 7
Costs
    infrastructure, 100
    transportation, 104
Countries, 52
    ASEAN, 7
Crisis, African
    urban, 76
Culture, 9

D
Dar es Salaam, 63
Database
    institutional, 48
    iterative, 50
    monumental, 48
    scientific, 119
Davidson, 36

Dawkins, 102
Dembele, 186
Design
    passive, 90
    urban, 28, 44
Desktop, 30
Determinism
    geographic, 101
    geographical, 49
Development
    isolated, 47
    leapfrog, 47
    low-density, 47
    project, 66
    ribbon or strip, 47
    scattered, 47
    sectoral, 99
Development planning, infrastructure, 142
Development theories, regional, 140
Dispensation, neocolonial, 143
Distribution
    dispersed, 53, 64
    of heterogeneous land, 53
Divisions
    territorial, 58
Doctrine
    of informality, 26
DRC, 30
Dynamics, urban, 101

E
Economic(s)
    developmental, 154
    dualism, 2
    geography, 3
    supply-side, 45
    survivalist informal sector, 46
Economic growth, regional, 140
Economic processes, 85
Economies, agglomeration, 121
Economists, 19
Economy
    African, 1–3
    market-space, 55
    space, 43
Education, planning, 66
Edwardo, 47
Egypt, 27
Entity, economic, 57
Environment
    regulatory, 75
    urban, 6

Index

Environmental management, urban, 131
Epistemology, 3, 4
  foundation, 1
  in planning, 1, 28
  of imperialism, 1, 5, 234
  of poverty, 1, 41
Equality, 82
ESDP, 127
Ethiopia, 30
Europe, 127
  mercantilism, 3
European, 26, 37
European union, 8
Evolution, path-dependent, 103
Expertise
  foreign, 62
  informal, 154
  local, 62
Expertise knowledge, formal, 201
External, 83
Externalities, development, 89

F

Facilities, educational, 90
Forces, market, 75
Form, urban
  base planning, 70
Foundation(s)
  active, 69
  epistemological, 77
Frameworks, planning, 66
Friedman, 16
Functional flow, 149, 150
Funding, 83
Funding, external, 136

G

Geography
  economic, 4, 10
Ghana, 132
Globalization, 2
  neoliberal, 50
Goods
  foreign, 48
Governance
  European, 128
  principles of, 177
  systems of, 175
  transaction, 107
Governance, good
  democratic, 162
  political, 162

Government
  state, 89
Growth, 1, 2, 7, 27, 35, 41, 45, 46, 48, 49, 91, 98, 104, 121, 122, 126, 130, 137, 141, 170, 182, 187, 196, 214, 235, 244
  economic, 2, 27, 43, 50, 51, 58, 73, 84, 89, 90, 100–104, 108, 110, 113, 115, 116, 120, 130, 151, 174, 181, 185, 196, 197, 199–201, 208, 209, 216, 242, 245
  endogenous, 6, 10, 50, 104, 108, 197
  exogenous, 103
  form-based, 76
  imperial, 177
  industrial, 87, 183
  integrated, 103
  national, 188
  nature of, 177
  neoclassical, 102
  of informalization, 37
  patterns of, 6
  population, 184, 222
  principle of, 109
  rapid, 59, 133, 229
  regional, 104–108, 197, 209
  rural, 237, 238, 240
  smart, 53, 58, 98, 119, 127, 205, 236
  spatial, 64, 206, 218, 235
  spatio-physical, 197
  structural, 58, 64, 206
  sustainable, 228
  traditional, 91
  transitional, 184
  urban, 53, 57, 58, 73, 117, 118, 120, 136, 138, 151, 184, 194, 198, 201, 206, 214, 222, 231, 232, 236, 238
Growth management, urban, 204
Growth rate, 63
  urban, 51
Growth triangle, 55
Guidelines, spatial, 201

H

Hawksley, 14
Hicks, 1, 2, 4, 28, 48, 65, 86, 187
Hongkong, 126
Housing, public
  rental, 90
Human, 82
Hypothesis
  convergence, 102

Hypothesis (*cont.*)
  divergence, 103
  short term, 133

I
Ideology, 15
IDP, 43
IDP initiative, definition of, 157
IDP practice, positive compliance of, 156
Imperialism, 5
Incremental, 113
Industries
  basic, 105
  domestic, 87, 183
  non-basic, 105
Informality, 26, 38
  definition of, 39
  of planning initiatives, 4
Informal sector, survivalist
  modern, 152
Infrastructure, information
  communication, 175
Initiative, of planning, 60
Innovations, aesthetic, 129
Institutional, 8
Institutions, local, 62
Institutions, political, 162
Instrument, analytical, 104
Instrument, rationale
  universal planning, 125
Instrumentality
  of imperialism, 12
Instruments, planning, 69
Integration, regional
  economic, 102
Integration
  economic bases of, 50
  spatial, 50
Interest, public, 112
International, 5
Interventions, sectoral, 177
Interview, personal
  interactive, 155
Investment, infrastructure, 104
Investment mechanisms, 38

J
Justice, distributional, 85
Justice
  spatial, 4

K
Kenya, 57
Keynesianism, social, 85, 182
  military, 85
Kingdoms
  African, 36
Knowledge, informal
  expert, 115
  of planning processes, 115
Krugman, 108

L
Labour, informal, 121
Land, 40
Landscape, planning, 130
Law, 40
Leadership, 69
  charismatic, 182
  visionary, 143
Legal, 8, 41
Levels, regional
  local, 66
Liberalism, 16
  classical, 26
Liberalization, trade, 86
Liberalization, 38
Literature
  critical, 91
  planning, 49
Location
  theories, 10

M
Machinery, planning, 58
Management
  land use, 9, 90
  sustainable, 145
Market economy, dualist, 138
Markets
  commodity, 3, 47
Master planning, formal, 152
Master planning, neoclassical, 128
Matlosana, 138
Measures, protective
  protectionist, 87
Mechanism, defence, 83
Mechanisms, as best practice, 117
Mercantilism
  Arab, 85
  Chinese, 15
  classical, 87
  Euro-American, 15
  European, 85
  Neo, 87
Mercantilist
  system, 84
Merchants

Arab, 35
Meta-theory, 16, 38, 91, 163
  neo-liberal, 92
  of planning, 92, 142, 162, 163
Mindset, 18
Mindset, positive, 186
Mission
  of cities, 31
Mode, technocentric, 76
Model
  development, 75
  hybrid, 99
  multi-nuclei, 122
  of planning, 44
  polycentric, 71
  survivalist, 120
Mohammadzadeh, 16, 91
Movement, anti-imperialist, 173
Movement, philosophical
  political, 177

N
Nation, egalitarian, 63
Nationalism, 13
Neoliberal, 2
Neo-Marxist, 2
NEPAD, 56
  initiative, 28
The Netherlands, 130
New economy, 10
Nigeria, 27, 132
Nodes, development, 67
None, 133
Normative, 113

O
Object
  of development, 40
Okeke, 11
Omenani, 81
Order, institutional, 103
Organization, spatial
  theory, 10
Organizations, community-based, 126
Organizations, economic
  of reciprocity, 96
  of residistribution, 96
  of the market, 96
Orientation, populist, 136
Orthodoxy
  economic, 11, 12, 88
Outlook, hegemonic, 88
Outlook, visionary, 186

P
Paradigm
  development, 18
  neo-African, 19
  planning, 19
  shift, 17
Participation, interactive
  is blueprint, 113
  public, 113
  target-oriented, 113
  typologies, 113
Participation, stakeholder, 125
Participatory planning, neoliberal, 136, 145
Partners
  development, 54
Patterns
  heterogeneous urban, 64
  homogeneous urban, 64
  nucleated, 121
  poly-nucleated, 122
Perception
  institutionalist, 61
  reformist, 55
Period, 68
  colonial, 37
  de-industrialization, 91
  historical, 14
  Keynesian, 25
  mercantilist, 25
Perspective(s)
  in planning, 28
  institutionalist, 44
  neo-classic planning, 110
  of informality, 44
  urban planning, 101
Philosophy, renaissance
  ethno-, 173
Philosophy
  neo-Marxist, 46
Plan instrument
  master, 8
Planner, 19
  African, 46
Planning, event-oriented
  sub-area development, 118
Planning
  advocacy, 113
  comprehensive, 116
  in a new master, 56
  land use, 44
  process-oriented, 8
  product-oriented, 8
  territorial, 6
  theory, 6

Planning Act, 127
Planning approach
  multi-sectoral, 156
  neo-comprehensive integrated, 128
  participatory, 152
  regulatory, 129
  spatial, 128
  traditional budgetary, 159
Planning concept, substantive, 199
Planning framework, pragmatic, 177
Planning initiative
  African, 141
  perception of, 161
  spatial, 130
Planning instrument, investment
  spatial, 159
Planning instrument
  spatial, 161
  weak, 155
  informal, 152
Planning methodology, 157
  spatial, 155, 157
Planning paradigm
  market-oriented, 152
  neo-liberal, 149
  neoliberal, 145
  new, 138
  participatory, 164
    formal, 159, 164
Planning perspective, high-tech, 210
Planning perspective, participatory, 162
Planning perspective, strategic
  participatory, 141
Planning perspectives, multi-dimensional
  new, 152
Planning practice
  basic scenario of, 162
  formal, 137
  statutory, 151
Planning principle, formal
  participatory, 149
Planning principle, spatial, 156
Planning principles, formal, 130
Planning process, 157
Planning processes, sectoral, 156
Planning rationality, the status of, 161
Planning rationality, 49
Planning region, formal
  economic, 140
  functional, 140
Planning schools, African, 245
Planning system, institutionalized, 126

Planning theory
  neoclassic, 29
  neoliberal, 129
  rational, 29
Plans, master, 72
Policy(ies)
  decentralization, 52, 64, 73, 121
  environmental, 73
Political scientists, 19
Politics, local, 104
Poor
  in diaspora, 41
Population, suburbanization of, 59
  of slum dwellers, 61
Population, 37
Potchefstroom, 20
Poverty, 48
Power
  imperial, 14
Practice, professional
  planning, 60
Principles
  neoliberalism, 15
Problems, environmental, 130
Procedure, planning, 86
Process, participatory
  planning, 74
Process(es)
  informalization, 50
  participatory, 42, 115
  visioning, 68
Processes, monitoring
  review, 113
Productivity, 1, 27, 37, 41, 51, 103, 116, 140,
    162, 183, 196, 199, 200, 207, 216,
    234
  effectively enhance, 5, 27, 28, 116, 169
  in Africa, 5
  negative, 27
  the declining urban, 48
  the per capita, 48, 51, 228
  urban, 202, 228
Professionals, external, 114
Profit
  creation of, 14
Profitability
  for private, 50, 83
  individual, 183
Programmes, urban upgrading, 62
Projects
  bankable, 135
  infrastructure, 89, 90

Index 267

SCP/EPM, 132
Protectionism, 84, 87

R
Rationality, 68
  in planning, 41
Region, polycentric
  metropolitan, 102
Region
  African, 12
  functional, 49
  urban economic, 49
Regional
  integration, 6
Regional development, integrated, 145, 169
Regionalism
  global, 84
  new, 30
Regions, core-periphery, 109
Regulation
  planning, 37
Relationships, unbalanced, 98
Renaissance
  Africa, 6
  African, 31, 92
Research, qualitative
  quantiative, 125
Research
  qualitative, 16
  quantitative, 16
Resilience
  of formal planning, 4
Resources
  natural, 12
Revolution
  cyclical, 110
  quiet, 30
Roy, 4
Rustenburg, 138

S
Schemes, housing, 98
Scholars
  neo-Marxists, 43
Science, regional, 105
SCP, 135
Sector
  economic, 48
  informal, 48, 72
  the public, 48
Sectoral planning, 157
Senegal, 57, 132
Services, consultancy, 72
Services, public, 108

Settlement, informal, 61
Settlement(s)
  Homeland, 35
  informal, 53
Singapore, 61
Slums, peripheral, 98
Socialism, 83
Society, 14, 82
  civil, 126
Sociologists, 19
South Africa, 16
Space, geographic, 122
Space
  economy, 2, 3
Space-activity, 7
Space economy, African, 142
Spatial planning, informal, 142
Spatial planning, 41
Spatial system, the development of, 163
Spatial systems, integrated, 152
Spatio-physical, 27
Sprawl
  urban, 47
Stakeholders
  a range of, 43
Standards, planning, 70
State
  City, 35
Statutory planning, resilience of
  broad guideline, 160
  dominance of, 160
Strategy
  informalization, 50
Structure
  deep, 11
  industrial, 104
  polycentric, 58
  report, 18
  superficial, 11
  urban, 98
Sustainability, urban, 127
System, planning
  human, 75
System
  historic, 84
  regulatory, 111
Systems analysis, world, 138

T
Tanzania, 27
Tendencies, 42
  manipulative, 115
Terminal, transit, 61
Territorial, 13

Territory, 9
  EU, 127
  geographic, 42
Theoretical framework, 152, 153, 159
  IDP, 156
  new, 154
  spatial planning, 156, 159
Theories, regional development
  institutionalist, 110
  neoclassical, 110
Theory, dependency
  modernization, 170
Theory, world-systems
  despendency, 95
Timeframe, good, 134
Timeframe, 12
Tlokwe, 138
Todes, 43, 129
Towns
  colonial, 37
  new, 37
Trade
  liberalization policies, 48
  retail, 48
Trade relations, international
  internal, 208
Trans-active, 113
Transitions
  from master planning, 62

U
Ubuntu, 81
Ujamaa, 81
UN-Habitat, 59
University
  North West, 20
Urban
  ecology, 11
  economy, 11
  form, 11
  productivity, 4, 5, 20, 27, 75, 139, 162
  space, 11
Urban form, western, 126
Urbanism, old
  modern, 151
  sustainable, 151
Urbanism, 64
  creative, 129
  new urbanism, 28

old urbanism, 28
sustainable, 133
traditional urbanism, 28
Urbanity
  core issue of, 46
Urbanization, 1, 2, 51, 53, 57, 59, 63, 64, 70, 73, 86, 136, 138, 139, 184, 190, 194, 206, 222, 233
  capitalist, 37
  city-based, 120
  colonial, 127
  contemporary, 138
  crisis of, 112
  dependent, 140
  dual, 58
  extroverted, 231
  increasing level of, 51
  introverted, 27, 64, 73, 235, 237
  metropolitan, 40
  pattern of, 26, 36, 121, 230
  poverty driven, 229
  process of, 112
  pseudo, 4, 46
  rapid, 86, 110, 162, 228, 233
  region-based, 120, 220
  socialist, 37
  sub, 47, 49, 53, 59, 74, 98, 121, 228, 235
  traditional, 36
Urban productivity, declining, 230
Urban regions, polycentric, 207

V
View, classical, 111
View, globalist
  pan-Africanist, 175
Vogue
  of, palnning, 30
Vulnerability, levels of, 183

W
Watson, 4
World Bank, 52–54
World-system, 16, 17

Z
Zambia, 132
Zones, economic, 57